T0302322

Understanding the Failure of Materials and Structures

Understanding the Failure of Materials and Structures introduces practical aspects of mechanical characterisation of materials and structures. It gives those with little or no prior experience insight into the process of developing everyday products, issues behind some high-profile failures, and tools to begin planning a programme of research.

Written in an easily accessible manner, the work discusses fundamentals of the physical world, highlighting the range of materials used and varied applications, and offers a brief history of materials development. It covers the role of materials structure in controlling materials properties and describes mechanical properties, such as stress, strain, stiffness, fracture, and fatigue. The book also features information on various modes of testing and strain measurement. It provides some discussion on topics that go beyond well-behaved test coupons, with thoughts on biomechanics, megastructures, and testing for applications in extreme environments. Finally, it covers how materials fail and the future of physical testing.

With minimal theory and mathematics, this work presents the fundamentals of mechanical characterisation of materials and structures in a manner accessible to the novice materials investigator and the layperson interested in the science behind materials engineered for use in common and advanced products.

David Jesson is a Materials Scientist by profession, training, and inclination: he is a Fellow of the Institute of Materials, Minerals, and Mining, and is both a Chartered Engineer and a Chartered Scientist. Following an undergraduate degree in Materials Science and Engineering from the University of Surrey, he undertook research leading to a PhD on the effects of nanoscale phases for the toughening of composite materials. On completion of his doctorate, he took a post as a Postdoctoral Research Fellow investigating the impact of in-service degradation on the performance of cast iron assets in the water industry, eventually leading into fifteen years of research and laboratory management. In 2021 he left academia to join the Materials Performance Group at Frazer-Nash Consultancy.

Understanding the Failure of Materials and Structures

An Introduction

David Jesson

CRC Press
Taylor & Francis Group
Boca Raton London New York

CRC Press is an imprint of the
Taylor & Francis Group, an **informa** business

Designed cover image: © David Jesson

First edition published 2025
by CRC Press
2385 NW Executive Center Drive, Suite 320, Boca Raton FL 33431

and by CRC Press
4 Park Square, Milton Park, Abingdon, Oxon, OX14 4RN

CRC Press is an imprint of Taylor & Francis Group, LLC

© 2025 David Jesson

ISBN: 978-0-367-36840-1 (hbk)
ISBN: 978-1-032-84166-3 (pbk)
ISBN: 978-0-367-82234-7 (ebk)

DOI: 10.1201/9780367822347

Typeset in Palatino
by Newgen Publishing UK

For my parents, without whom I could not have started this book,

And for Ellie, without whom I could not have finished it.

Contents

Preface

At the time of writing this preface, one of the very last parts of the book to be written, several announcements have been made about new materials, including "High fatigue resistance in a titanium alloy via near-void-free 3D printing", which was published in *Nature*, on the 28th February, 2024. As someone who has broken a wide variety of specimens whilst investigating materials for automotive, aerospace, defence, and civil engineering applications, I find the testing of new materials every bit as fascinating as the new materials themselves.

As well as undertaking my own research, I have also managed a laboratory, and have therefore had the exciting opportunity of supporting students as they study new materials, or new applications. Whether undergraduate or postgraduate, engineer, physicist, chemist, the vast majority had one thing in common: this was their first experience of mechanical characterisation. Some had a theoretical understanding of one aspect or another. Some arrived gripping a standard. But no-one arrived knowing exactly what they needed to do, or the implications of what their test required.

This book was born from a number of repeated conversations: engineers usually understand the mechanics of the test, but very little about the materials and what effect their fundamental nature has on the test, whilst physicists and chemists usually understand the materials but need a helping hand to get started with the testing.

I cannot claim to have captured every single aspect of mechanical characterisation, but I have done my best to lay out the fundamentals for undertaking the mechanical characterisation of materials and the impact of microstructure on testing. I hope those coming to mechanical characterisation for the first time will find the book of use as a primer, and those coming back to testing will find something in these pages which will make their testing easier.

Happy researching!
David Jesson,
March 2024

Acknowledgements

No book is published without considerable effort, not all of which is down to the author.

It has been a great privilege to teach and supervise students at all levels: I would like to thank those students for all that they have taught me.

My huge thanks to Allison Shatkin of Taylor and Francis for her faith in accepting the proposal for this book, and her ongoing support whilst I was writing it.

Dr Steve Martin, educator *extraordinaire*, introduced me to Materials Science all those years ago when I was studying Physics in preparation for university. I'd known that I wanted to study Materials Science for nearly ten years before that, I just didn't know what it was. Dr Martin set my feet on the right path, not to mention demonstrating the threat of finely divided particles placed in a cloud near a flame.

There are four people who have had an enormous impact upon my professional life, and in some respects set my feet onto the path that has led to the writing of this book. Professors Paul Smith and John Watts accepted my application to do a PhD and supervised my first research efforts, some of which underpin this book. Dr Mike Mulheron has been a fount of knowledge and a pillar of support as I attempted to serve two masters – research and laboratory management. Last but by no means least, Mr Peter Haynes, now retired, has been a constant source of practical knowledge, enthusiasm, and support for more years than either of us care to remember. There are many excellent reasons why I changed some of the computer screensavers in the Mechanical Testing Facility at the University of Surrey to read "Always do what Pete says".

1

Learning from Failure

We learn from failure, not from success. – Bram Stoker

All my successes have been built on my failures. – Benjamin Disraeli

1.1 Introduction

On July 23rd, 1983, Air Canada Flight 143 was forced to make an emergency landing. The flight became known as the Gimli Glider – not for the dwarfish member of the Fellowship of the Ring, but because the pilot effected an emergency glider landing at a former Royal Canadian Airforce base at Gimli, Manitoba. The cause of the emergency is commonly stated to be a mistake in fuelling caused by confusing metric and imperial units, leading to the plane carrying less than 50% of the required fuel (e.g., Witkin, 1983). In fact, the cause of the accident was much more complex than this simple error – the fuel issue was just the last in a number of incidents, leading to what is termed a 'Swiss cheese failure'[1] (Klein, 2011). The outcome of this event, happily, resulted in no loss of life: the pilot was skilled and was able to land the plane despite a lack of emergency protocols for the situation he found himself in. Unusually, he had experience as a glider pilot, and this stood him in good stead. The 'Swiss cheese failure' could have ended in tragedy if a few more factors had lined up.

Fundamentally though, whilst a faulty gauge was at the heart of the problem, the failure was one of human error. People made calculation errors, because they used incorrect conversion factors. A technician was distracted and forgot to turn an indicator off. The aircraft structure, however, was sound: the failure in this instance was fundamentally procedural.

---

[1] A 'Swiss cheese failure' is one where all the holes, which is to say the things that could go wrong, line up, and give rise to an event that leads to failure. It might not matter that a draught causes a door to slam, but if that door is your front door and you don't have your keys, you will be locked out.

DOI: 10.1201/9780367822347-1

Another well-documented failure is that of the Apollo XIII mission to the Moon. Here the cause of the explosion which aborted the mission[2] was shown to be due to a combination of damaged Teflon insulation and materials that were not completely stable when in contact with supercritical oxygen (United States, 1970). The cause of the damage to the insulation was not determined. Here then, whilst the inciting incident was procedural – an appropriate request to stir the tanks to determine if a sensor was malfunctioning – the failure was caused by prior damage to the assembly. As with the Gimli Glider though, the eventual outcome is seen as something of a triumph: even though their mission was aborted, the crew of three were returned to Earth safely.

There have been other events though which rightly carry the label 'disaster'. For example, the Tay Bridge disaster (28th December 1879) saw an estimated 75 people killed when the railway bridge across the River Tay in Scotland collapsed during a period of high winds measuring 10–11 on the Beaufort Scale. The enquiry into the disaster noted several issues, but the panel struggled to agree an overall cause for the event. Contributing factors included the fundamental design, the quality of materials used, and the speed with which some trains crossed the bridge (Court of Enquiry, 1880). Modern re-examinations of the evidence are popular. Jones (1994) tends to the view that the design, which did not take into account wind-loading, was the critical issue, whilst Martin and Macleod (1995) take a more holistic view of the design, noting that the architect was forced to make some significant changes to the design due to the ground to be built on not having the depth of rock originally thought to be present, and due to the spiralling costs of the project. What was not well understood at the time, but which came to light subsequently, was that cast iron, a major part of the bridge, behaves differently in tension compared with compression. The compressive strength is much greater than the tensile strength. Further, cast iron contains graphite flakes which can act as 'seeds' for crack growth. The oscillations reported by one of the witnesses may well have been sufficient to cause fatigue cracks to initiate unobserved within the structural elements and Lewis and Reynolds (2002) note that this was almost certainly a contributing factor.[3]

Last, but by no means least in this introduction to failure, we have the de Havilland Comet. The world's first commercial airliner suffered a few high-profile crashes in the early 1950s, after having been in successful service for some time. Two early failures were successfully investigated, and problems resolved. But then several more occurred. The problem was so severe, and so inexplicable, that the aircraft class's airworthiness certificate was revoked, and entire fleets grounded and moth-balled until the problem was resolved. New designs went into manufacture, but it didn't resolve the problems

---

[2] "Houston, we've had a problem." – James Lovell
[3] They also comment on the nature of the investigation, one of the first systematic inquiries of its kind, and probably pioneering in its use of photography to log certain kinds of evidence.

associated with the Class I. Eventually it was discovered that the square-framed windows were acting to concentrate the stress, leading to the development of fatigue cracks.

Setting aside human error, failures are a result of the action of the environment on the structure of interest, where such action exceeds the material's ability to cope. By causing failure on our own terms, by examining failures that occur unexpectedly, we can collect data that inform models and decision-making processes. If we are careful, we can design better, safer, longer-lasting structures – if that is our focus. Or we can design materials that will fail in a manner of our choosing, for example biodegradable packaging, safety glass, and props for stunts: in these instances, we need to ensure that the item has sufficient strength for the job it is to be used for, but that it will fail in the right way at the right time.

1.2 What Do We Mean by Failure?

When dealing with inspirational quotes, it is perhaps traditional, especially in an engineering setting, to start with Thomas Edison's "I have not failed. I have just found ten thousand ways that won't work". Equally, we might take Henry Ford's maxim: "Failure is simply the opportunity to begin again, this time more intelligently". Instead, this chapter begins with the musings of an actor and of a politician; musings that are similar but subtly different. They are of course talking about their various endeavours, but if we wilfully misunderstand them and take failure in an engineering sense, then we can discern the fundamental reason for breaking things: to understand the potential mechanical performance of materials and structures. But what do we mean by 'failure'?

At this point it is traditional to give a dictionary definition: The Oxford English Dictionary gives six definitions of the word failure:

1. lack of success; failing
2. an unsuccessful person, thing, or attempt
3. non-performance, non-occurrence
4. breaking down or ceasing to function (heart failure, engine failure)
5. running short of supply, etc.
6. bankruptcy, collapse

The trend is clear, but only #4 is appropriate for the purposes of this book, and even this is too limited. From an engineering perspective, failure is as simple as the thing of interest being unable to do the job for which it is designed, but failure can look like the sinking of a ship ('unsinkable' or otherwise), a cave-in in a mine, faulty items coming off a production line, or a great many other things, including seemingly trivial things like a liquid leaking from a failed seal.[4] The thing in question has ceased to function, it has failed: but simply saying that it has failed is seldom enough. We need to understand why it has failed.

Let us consider the timeline of a failure:

Inciting incident(s) → Event → Aftermath

The event itself is typically a moment of time which might be as short as a few microseconds or potentially as long as a few moments.[5] It is the point at which the key characteristic of the item is overwhelmed by the situation it finds itself in. A defective item can survive a surprisingly long time, provided it is not subjected to a critical event. A wooden bench, for example, might be rotten to the point of instability, but will only fail when someone sits on it.

The aftermath is the province initially of those who must deal with whatever happened. This might be as simple as mopping the floor but might require emergency services. Once all is made safe, investigators will determine what happened. This might be as simple as someone saying "Well, I won't pick up a can with oily hands again" or it might require the painstaking collection of every single piece of an aircraft and months of examinations.

During the aftermath, as well as looking at the physical evidence of the broken item, the investigators will take a history, looking at the factors that might have contributed to the failure. This history may be very straightforward or may take in multiple contributing factors; it may be of the order of seconds, minutes, or days between an inciting incident and the failure itself, or it may be years, decades, or centuries. The timescale will be dependent on the mechanisms at work.

Failure does not have to be catastrophic of course. A worn item can fail and stop doing its job. This might happen if a manufacturing process miscues, for

---

[4] The consequences of such a leakage can seem trivial if the liquid in question is water, and more obviously problematic if the liquid is oil or brake fluid, for example. However, it depends on how widely we are thinking. On the one hand, given the over-consumption of water in the West and the potential for water-stress, water wasted through leaks has some serious socio-economic considerations. On the other hand, a leak underground, going unobserved for many years can give rise to perfect conditions for accelerated corrosion, and/or a form of in-soil turbulence which can cause pebbles to rotate and wear a deeper hole in the pipe, leading to a catastrophic failure in due course. Hence, even relatively minor 'failures' should be prevented or addressed in a timely fashion. A trunk main burst, for example, can cause the release of millions of litres of water, untold water damage to local properties, disruption, and the shutting off of the supply to local householders.

[5] Where a moment was historically quantified as a fortieth of a solar hour, or about 90 seconds, depending on the time of day and the time of the year. If this sounds imprecise, it is: it's a function of attempting to determine accuracy with inaccurate tools, at a time when 99% of the population measured time by religious ritual, meal times, and whether there was enough natural light to work or not.

example, and a component ends up misshapen: it is not capable of fulfilling its designed purpose. Or if, perhaps, the moulds or dies used in the manufacturing operation have become worn or damaged through operation, or detailing has become filled with detritus or lubrication.[6]

Further, this engineering concept of failure allows us to see that it is possible to design things so that should it occur, the structure will fail in a manner by which least harm should occur. An obvious example is a pressure vessel, which normally includes some kind of relief valve, should the pressure increase above a defined level, perhaps in the event of a fire. A soft failure is one in which rupture occurs, and the contents spill, but in such a way as to prevent an explosion, or something similar due to the instantaneous release of strain energy arising from the load placed on the structure. A safe failure is one in which the failure of a component causes the system to stop operating, without causing problems elsewhere in the system.

When investigating failure, it is much like carrying out a post-mortem on a corpse. Instead of saying that the subject died of heart failure, we might say that the specimen failed because it could not stand the applied stress. But this is as unsatisfactory as stating heart failure. The heart stopped and the person died: but why did the heart stop? Was it old and tired? Did the person ingest something they shouldn't have? Was the heart over-stressed because the person was exercising too hard, or had become terrified of something, or …? Similarly, if a structure has become over-stressed, is that a design fault, an operational one, or has the strength of the structure become degraded for some reason?

1.3 Why, Then, Do Materials and Structures Fail?

Which leaves us with the question of why do materials and structures fail? In the context of this book, we can leave aside questions of poor design or operational over-loading. Instead, let us focus on the behaviour of materials. There are a number of mechanisms which will lead to degradation of a structure and ultimately failure. These include:

- Wear: two parts moving against each other, giving rise to damage, e.g., a rope in contact with a hard surface; gears in a mechanism; footfall.[7]

---

[6] In some instances, such failures can become more valuable than the original item, such as in the case of coins and stamps where striking and printing errors give rise to rare defects.

[7] In this instance it is more likely to be the soles of the shoes that become worn, especially in the short term, but consider stone steps in an old building and the distinctive dip that occurs from generations of feet passing up and down, for example.

- Chemical attack: often oxidation giving rise to corrosion, but other chemicals can give rise to other mechanisms.[8] Degradation can be generalised over the whole surface or can give lead to pin-hole failures.
- Microbial attack: like chemical attack, although there is rarely a protective aspect, and the overall mechanism may essentially be a form of chemical attack, although the ultimate cause arises from the presence of certain bacteria.
- Insect or animal attack: this can be thought of as a form of wear, in some respects, but is generally more active.
- Formation of cracks from trauma or cyclic loading.
- Electromagnetic radiation: some wavelengths of energy are particularly problematic and can cause damage at a microstructural level. For example, polymers are particularly susceptible to bond damage arising from the UV portion of the spectrum.

Ultimately, failure of structures requires an understanding of materials and the manner in which they fail, and this could be summed up as change in the relationship between the constituent atoms, such that the material no longer has the properties that we thought it had. This might be due to atoms being removed or chemical bonds being changed.

It is worth noting that sometimes 'failure' can work to our advantage. Some materials, most commonly metals, but also including some polymers, can be made stronger through a process called 'work-hardening', sometimes referred to as 'strain-hardening'. Here, defects within the crystal structure move about when a load is applied. They interact, and become locked in place, meaning that the next time that a load is applied, a much larger stress is required to cause further deformation. Technically, the material has failed: we have loaded the material beyond its elastic limit and caused plastic deformation to occur. In the process we have made the material stronger. Under some circumstances however, this strength can be a problem. But how can that be? If I have designed for a certain strength and my material is stronger, this must be good? Unfortunately, this increase in strength generally comes at the expense of ductility: there is a loss of toughness, and the material becomes more brittle. Hence, whilst work-hardening can be beneficial, and is actively pursued, in other instances it is a nuisance that must be prevented or mitigated.

---

[8] Some chemical attacks can be self-limiting, leading to passivation and subsequent protective coatings. Stainless steel, for example, is protected by the thick oxide layer which forms on the surface and prevents further oxidation from occurring, unless the layer is damaged. Unless treated, the layer on aluminium is porous and so attack can continue, although a small quantity of aluminium added to silver leads to a passivating oxide layer which prevents tarnish from occurring. Protective patinas of various copper-based minerals can be formed on copper.

More usually, failure is the consequence of degradation. Whilst this is something that begins almost as soon as the item in question is produced, the rate at which it occurs can be variable, and sometimes multiple mechanisms can be in play; such mechanisms can work independently or can reinforce one another. For example, consider the Great Pyramid at Giza. This was built more than 4500 years ago and is still recognisable as a great monument. At different times it has been affected by scour, damage arising from earthquakes, pollution, and changes in the water table (Hemeda and Sonbol, 2020). It is difficult to know whether the Great Pyramid will still exist in any meaningful form in another 4500 years without significant investment.

There are all manner of degradation processes. Oxide films that form on metals: in some cases, these can be protective or can be enhanced to make them so; in others they are permeable, and the metal continues to oxidise. Rain works its way into small cracks: the temperature drops, the water freezes, and the crack is levered open. Insects and microbes start eating away at the softer tissues in a piece of wood, hollowing it out in a seen or unseen manner.

Hence, whilst failure may be due to a sudden trauma, such as an out-of-control car hitting a fence, or a wall, or a tree, some failures are years in the making as defects are created and grow until they become critical.

1.4 Purpose and Outline

Understanding the failure of materials and structures is essential if we are to prevent catastrophe when accidents happen. Understanding the nature of a failure, when it occurs, is essential to preventing the same thing happening again. Whilst failure may come from a short-circuit causing over-heating, or a spark, leading to a fire, or some other process, these are generally specific issues. A fundamental aspect of any structure is that the material it is made from must be at least strong enough to hold its own weight: the vast majority of design effort is ensuring that the structure can deal with normal and abnormal operating conditions – such as the possibility of variable wind loading on a bridge.

Different disciplines will come to mechanical testing at different times to get information about the behaviour of materials and structures. For some, mechanical characterisation is the bedrock of their practice. For others, it is a dalliance that comes occasionally, to provide a specific piece of information for a single purpose. Hence, this book has come about through 20 years' experience of interdisciplinary research and laboratory management. It is an encapsulation of training talks, discussions about the best way to test certain materials, and the best way to analyse the resulting data.

The fundamental physical principles underlying mechanical properties are presented, including the impact of defects, and the effects of repeated low-level loading. To some, some ideas presented will be extremely obvious, and in certain instances only the most basic of ideas are included. This book is an introduction; it is not a textbook on chemistry or physics or engineering but will be of use to all these disciplines and more. Hence, there are certain fundamentals that are essential to a good grounding in the subject, but further reading will almost certainly be necessary to unpick a more detailed point.

In terms of the practicalities of mechanical characterisation, modes of testing and their relationship with mechanical properties are discussed, together with a separate chapter on the means of measuring strain and the complex issue of machine compliance. This chapter considers both the traditional and more modern methods of strain measurement.

Testing can and should be straightforward, for the most part: there will always be cases where complexity arises. However, there is, in a sense, a corollary to Heisenberg's Uncertainty when testing: how can we be sure that our testing has not affected the sample that we are testing? There are some obvious steps to be taken to minimise such problems, one of which is to be sure one understands the issues around strain-rate.

Frequently, failure is reduced to ductile or brittle failure, which owes much to the influence of metallurgy on the development of Materials Science. It is important to understand the mechanisms that give rise to these sorts of failure but also to understand what other kinds of failure might occur, particularly those associated with other materials. For example, composite materials give rise to a range of failure modes not seen elsewhere.

This book also gives some thought to the use of statistics when considering failure. Some failures such as the Gimli Glider are, or should be, unique. Others sit in a spectrum of failure: lightbulbs for example are generally expected to have a certain lifetime, but they rarely fail at that stated lifetime. If you were to install a set of brand-new lightbulbs around your home, it would be unusual in the extreme if they all failed at the same moment, even if you ran them all continuously.[9] Statistics can help us understand failure in several ways.

Remarks on understanding failure through the use of models are also presented: there is, quite rightly, a growing use of finite element modelling and similar processes to design structures, set design limits, and predict failure. Such methods are essential to innovation if design costs are to be kept to a minimum. The problem comes when such models are relied on, and absolute belief in their output trumps real-world test data. Test and model need to agree if the model is to be useful.

---

[9] Barring, of course, some unfortunate event such as a power-surge that killed all the bulbs simultaneously.

The book finishes with a look at some of the challenges faced by some extreme situations, and with some thoughts on 'unbreakable' materials and the structures of the future.

Rather than gather all references at the end of the book, those for a specific chapter follow on from a summary. Some, but not all, chapters also include some recommended reading, which whilst not explicitly referenced in the text has informed the book and is considered useful for anyone looking to improve their understanding of materials, their mechanical characterisation, their engineering uses, and ultimately their failure.

References

Court of Enquiry. Report upon the circumstances attending the fall of a portion of the Tay Bridge, 1880. (See e.g., www.railwaysarchive.co.uk/documents/BoT_Tay Inquiry1880.pdf, accessed 07/06/21)

Hemeda, S. and Sonbol, A., 2020. Sustainability problems of the Giza pyramids. *Heritage Science*, 8(1), pp.8.

Jones, D.R.H., 1994. The Tay bridge disaster—Faulty materials or faulty design?. *Engineering Failure Analysis*, 1(3), pp.243–253.

Klein, G.A., 2011. *Streetlights and shadows: Searching for the keys to adaptive decision making*. MIT Press.

Lewis, P.M.R. and Reynolds, K., 2002. Forensic engineering: A reappraisal of the Tay Bridge disaster. *Interdisciplinary Science Reviews*, 27(4), pp.287–298.

Martin, T. and Macleod, I.A., 1995, May. The Tay Rail Bridge Disaster-A Reappraisal Based on Modern Analysis Methods. In *Proceedings of the Institution of Civil Engineers-Civil Engineering* (Vol. 108, No. 2, pp. 77–83). Thomas Telford-ICE Virtual Library.

United States. National Aeronautics and Space Administration and Cortright, E.M., 1970. Report of Apollo 13 review board. National Aeronautics and Space Administration.

Witkin, R., 1983. Jet's fuel ran out after metric conversion errors. *New York Times*.

2

Why Break Things?

Stress analysis is an important part of engineering science, as failure of most engineering components is usually due to stress. The component under stress investigation can vary from the legs of an integrated circuit to the legs of an off-shore drilling rig, or from a submarine pressure hull to the fuselage of a jumbo jet engine.

– Strength of Materials and Structures, 4th Ed. Case, Chilver, and Ross

2.1 The Material World

Our world is shaped, literally and figuratively, by materials. Ever since our distant ancestors learned to use tools[1] we have been changing ourselves, the landscape, and our environment. Sticks and stones have been joined by metals, plastics, and ceramics, together with larger scale timber structures and synthetic stone in the form of concrete. Through trial and error we learned that not all wood is the same: there is wood which is better for building, for making tools with, and wood that is just better burnt. Timber can be a material just by being cut and shaped, or it can be processed to give us various chemicals, and of course paper. Sometimes it does not even need to be cut down to provide useful products. A number of trees can be tapped, and their saps used for various applications, ranging from the sugary saps of maple and birch through to the latexes used to produce rubber, which are derived from trees such as *Hevea brasiliensis*, for example. Through trial and error, we learned to shape flint to give sharp edges for different applications – axes, knives, arrowheads.

---

[1] To date, the oldest known cache of tools, indicating the use of both stone and wood, were found in Schöningen, Germany. Originally believed to have been up to 400,000 years old, a recent assessment (Richter and Krbetschek, 2015) suggests that they are as recent as 300,000. In either case they are associated with H. *heidelbergensis* rather than *H. sapiens*, i.e., an extinct ancestor to the humans now inhabiting the Earth.

DOI: 10.1201/9780367822347-2

Today as you read this book, unless you have gone to some special effort to remove yourself from the daily routine, you are surrounded by all sorts of materials: fabrics for clothing and furnishings; timber, brick, concrete, and other construction materials; metals for wiring, cookware, white ware; paper for books, letters, packaging; plastics – everywhere!; and of course the dense concentration of a plethora of different materials used in the range of electronic devices.[2]

All these materials serve a purpose; each represents a considered selection of a set of properties. Materials may be chosen for a property other than their strength, although this may well be a secondary consideration in such cases. The suite of properties that might be important for a design include optical, electrical, and corrosion resistance, to name but three.

But, the critical issue at this point, is that no one considers a material for its own sake. We choose materials because we want to make things: bridges, cars, aircraft, spacecraft, phones – the list is that of life, ranging from everyday existence through to the challenges that we set ourselves as a species, and not forgetting artistic expression. Every structure that we make is dependent on design choices, and every design choice is dependent on an understanding of materials and what they can do for us.

People have been observing the natural world since before we could really be considered people. Our primitive ancestors would have behaved much as the creatures we see around us today: they would have looked for telltale signs of predators, for food, and for changes in the environment that would have hinted at short-term weather patterns. Those that were good at this survived and passed on their knowledge; those that were less skilled did not. This is of course an over-generalisation, but since we are not considering the evolution of species, we will leave the point there to focus on the evolution of scientists and engineers.

In *Guns, Germs, and Steel*, Diamond (1998) summarises the evolution of societies, amongst other factors, to explain why the range of societies that we have today exists in the way that it does. One factor that is considered in detail is the rise of farming, at the expense of the hunter-gatherer lifestyle. Whilst there are a surprising number of disadvantages to sticking in one spot, one of the eventual advantages is that instead of the entire band (typically 10–20 people comprising one or two extended families), several bands can come together as a tribe (hundreds of people) or eventually a chiefdom (thousands), and a proportion of the people can grow all the food for everybody. This, of course, gives rise to all sorts of issues that are the domain of politicians, social scientists, and anthropologists: a class structure, bureaucracy, specialists. It

---

[2] There are around forty different elements used in a mobile phone, for example (UNEP, 2009; OECD, 2010). The actual number of materials is harder to assess because there are several different plastics (which are, in general, variations on the theme of carbon and hydrogen, sometimes with oxygen, nitrogen, and/or one or other halogens). On the other hand, several elements form unique alloys.

is the last that we are most interested in: specialists do not have to spend their days collecting food to live (unless they are the specialist farmers, of course), and can therefore develop their skills, learning to improve on the basics. Amongst other things, this evolution of society gives the specialists more time to think, and to undertake experiments.

At this point, everyone was a scientist in some respect, but there were no specialist scientists. The farmers were undertaking genetic breeding experiments, of a sort, the flint knappers were learning to discriminate between good- and inferior-quality flints. Potters were learning to mould clay; metal workers were discovering how to work copper, bronze, iron. Leaders were learning how to make use of the workforce available to embark on civic works.

2.2 Specialist

In modern times, it might be argued that we are drifting into a state of over-specialisation at too early a stage. Robert Heinlein once said that "specialization was for insects": he felt that people ought to be able to take responsibility for everything that touched their lives, ranging from building your own house through to growing and cooking your own food, and beyond. The problem with this is that there are so many things which work better with more sophisticated tools and machinery than are available in the home environment. So, for example, it would be possible to make the paper for (the physical edition of) this book at home, but it would not be as smooth as the paper manufactured in a paper-mill.

In terms of specialisation, the role of the scientist as we would understand it today is a recent construct when compared with the emerging specialists of pre-history. It is accepted that it was not until the 19th century that the professional scientist really came into being.[3] This is not to say that scientists did not exist before this, but they tended to be hobbyists, often working on their experiments part time whilst employed as government officials or teachers, for example. The earliest people we think of as scientists are usually referred to as philosophers[4] and were simultaneously engaged at several points along a spectrum that takes in the various disciplines (including subjects now discredited, such as astrology) and engineering. For example, Imhotep, the

---

[3] Ironically, it was one of the last great generalists, William Whewell, who proposed the term 'scientist' in 1833 to overcome issues with descriptions. By this time, specialisation was becoming the norm, with increasingly niche areas of focus becoming at once popular and derided. At this point, a university education was still something of a rarity and one of the greatest professional scientists of the time, Michael Faraday, had no formal education in this regard, but rather served his time as a laboratory assistant and learned his craft in this manner.

[4] Literally, lovers (philo) of knowledge (sofia).

builder of the Step Pyramid (built between approximately 2630 and 2611 BC), was an important government official and is credited with being the architect of this early pyramid. Contemporaneous evidence is scarce, but he was clearly an influential figure: after his death he was deified and credited with the writing of various wisdom texts and with being a great physician. Several thousand years later, and the pattern continues with people such as Archimedes, who was clearly doing some deep thinking about statics and hydrostatics, and equally clearly was making 'engines'. Again, looking at a body of work from a distance in time, it can be difficult to determine exactly what he was responsible for inventing, what he improved upon, and what has been corrupted down the centuries. It is interesting to note, however, that he is supposed to have been most proud of his proof that both the volume and the surface area of a sphere are two-thirds that of a cylinder (including the ends), where the height of the cylinder is equal to the diameter of the sphere.

By contrast, the concept of a professional engineer is much older. The word engineer dates to the Middle Ages, and is associated particularly with the building of fortifications, but a similar role can be found in the Roman military. The concept of an engineer was also applied in civil applications, in terms of the construction of large buildings, naval architecture, and in areas that today we would consider to be the domain of mechanical engineering.

Fundamentally, the learning point that can be taken from this is that there is a great overlap between the function of the engineer and that of the scientist, particularly in certain areas of study. One of the things that we associate more closely with engineering rather than science, however, is the process of design.

2.3 The Design Round

The process of designing an object can be incredibly complex. A car, for example, can take several years to go from a concept to the point that it is rolling off the production line, and this speed is usually associated with a refinement or update to an existing model: a completely new design may be five years in development, depending on the number of completely new components that need to be tested. Similarly, an aircraft may be developed from an entirely new concept or may be based on a previous model. The iconic 747, for example, owes some of its design to previous developments for other applications, including Boeing's bid for the military CX-HLS (Heavy Logistics System). Even so, it is surprising to consider that the 747 project took only 28 months to go from design to test flights; the first aircraft being delivered to the now-defunct Pan Am and in commercial use by

1970. Equally surprising is that less than 1600 planes based on the 747 design have ever been produced. There are several variants, including the VC-25, better known by its call-sign of Airforce One, the specially fitted version used to transport the US president and entourage, and the E4 variant used as a mobile command post by the US Government.

The design process is invariably complicated and fraught with last-minute changes. An extreme example, presented anecdotally, was that of a container to hold reaction mass[5] for a satellite. Payloads for spacecraft are limited by both volume (in terms of shape as well as total space available) and mass. Equally though, a launch is expensive, and it makes sense to maximise the usage of the available space. But if the reason that you are undertaking the launch in the first place is to put a communications satellite weighing several tons into orbit, the left-over space can be oddly shaped. Still, you want to make the most of the pocket available, and many scientific experiments hefted into orbit do not need to be beautiful, or even to last exceptionally long, but they do need to have specific features. In this example, a 'nano-sat'[6] was taking advantage of some dead space that was present in the cargo hold of a launch: therefore, there was a specific volume available, both in terms of shape and size. The focus was on the mission-critical features of the satellite – instrumentation and the like. The remaining space for the fuel tank was, surprisingly, triangular in cross-section. This is a less than ideal shape to work with, because under pressure, the triangular corners will be weak points. Happily, it was not necessary to use the entire volume for the reaction mass required for the mission, and this enabled some creative design approaches to be considered. The final design was for some piping to be coiled into a shape that might be described as a triangular spring.

There are therefore several facets to the design round, which include:

I.   The purpose of the overall structure or device. A car, for example, will require distinct characteristics dependent on whether it is a sports car, designed to go fast, or a family car, designed to carry everything and the kitchen sink. Similarly, if we consider a laptop, an ultralight device is unlikely to have much capacity in terms of integral memory or battery

---

[5] Reaction mass is used to manoeuvre a satellite in space. The mass is not burned as in a combustion engine but is typically pressurised, and sometimes additionally heated by electricity produced from solar panels on the satellite. The reaction mass is then released from various nozzles placed around the satellite to move it. The usual reason for this is to return it to its orbit when the orbit degrades over time and gravitational attraction of the Earth starts to draw it home. In some instances, it might be necessary to move the satellite's position relative to the Earth, but in reaction mass terms this can be expensive. Some satellites can be 'refuelled', but this can be dangerous and not entirely straightforward, so in many cases once the reaction mass has been used up, this marks the end phase for the satellite. It may still have many useful years ahead of it, but its functionality has been reduced because it can no longer be moved.
[6] A satellite massing only a few kilograms.

capacity. On the other hand, there are laptops that might be described as 'luggable' rather than portable and these tend to have higher performance such as in the case of a gaming computer which will have power- and memory-hungry dedicated graphics cards on board. Then again, a laptop might be designed with robustness in mind, being required for use in difficult field conditions. Such toughness might be achieved by using a flight case, or similar hard-skinned, reinforced box, but these can reduce the usability of the device. Further, whilst the laptop is removable from the bulky protection, this is not always desirable. Whilst obvious, having removed the protection, the device is no longer protected. The resilience required might be achieved more permanently using a more robust (but expensive) electronic architecture and using a more rugged casing.

II. The interaction between various subsystems. Most things that are made are formed of more than one component. Some are passive, for example a protective coating. In other situations, there is a range of systems that need to work together either directly or indirectly. Returning to our automotive example, working from the prime functionality of a device to move from A to B carrying passengers and cargo, we can see that there is something that needs to move and something that needs to provide power to enable that movement to occur. The wheels move, so they need to be attached to the chassis in such a way that they can rotate, and there needs to be a linkage between the engine and the wheels to transfer the power generated to where it needs to be. This linkage may be simple or complicated, depending on whether the car is four-wheel drive, rear-wheel drive, or front-wheel drive. But, one has only to look at a modern car in comparison to an early one to know that things have become more complex. For example, to start cars one used to have to crank a starting handle to initiate the combustion process by providing the initial compression cycle – get this wrong and it was all too easy to end up with a broken wrist or arm. This has evolved to the electric self-starter that we know and love today – but this then requires an electrical system. On the other hand, this electric system can also be used to replace candle or oil lamps, and these days to power the entertainment systems that have superseded simple radios, and GPS navigation.

III. Safety requirements, which may be dependent on location, or overall function, and which may need to consider bystanders as much as the user. Again, the automotive industry provides obvious examples. Bonnet ornaments were required to be removed to protect pedestrians in the event of a collision, and indeed there have also been requirements to change the shape of cars so that in the event of a collision a pedestrian is not flung up in such a way as to cause greater injury. Further, whilst advances in materials have led to cars becoming lighter, increases in safety features, navigation systems, and entertainment suites have led to cars becoming heavier (MacKenzie et al., 2014).

IV. Other factors, including improving functionality/usability, company-specific aesthetics, and, increasingly, sustainability features such as Nike's use of recycled plastics in its product range.

2.4 The Pyramid of Tests

If you start with megalithic stone structures, humans have been building things in a meaningful way for at least five thousand years. In a mere six hundred or so years we go from 'simple' structures, with one stone lintel placed on top of two uprights, such as Stonehenge, to the pyramids of Egypt, and on to the large-scale structures of ancient China and Mesoamerica.[7] These were built with the materials available and were designed based on tried and tested methods: empirical design. Empirical methods were still employed during Europe's Medieval period when many of the great cathedrals and castles were built, sometimes over the course of hundreds of years.

But it is only within the last hundred or so years that we have moved towards predictive design, based on models, and it is only in the last fifty or so years that this design process has been dependent on the materials available, especially those of a synthetic nature.

The Pyramid of Tests sets out a hierarchy of testing which starts with materials at the bottom and finishes with structures at the top (Figure 2.1).[8] The most frequently cited example of a hierarchical test program is the Rouchon Pyramid (Rouchon, 1990), which arose from one specific situation: the intention to use composite materials in aerospace structures. Whilst metals were well understood, and could be formed easily into the required components, and heat treated and aged to get a specific set of properties, composite materials were still something of an unknown and there was a significant process required to validate parts for use. Hence, Rouchon (1990), specifically considering composite materials, has coupons formed from resins and fibres at the base of his pyramid and this simplest testing includes consideration of

---

[7] Although it is unlikely in the extreme that there was sufficient exchange of ideas for these structures to influence the others.

[8] Jerry Lord of Boeing has suggested that there should be a second pyramid that should be considered at the same time. Whilst not entirely relevant to the current context, it is worth noting for completeness. This second pyramid considers the validation of the manufacturing process.

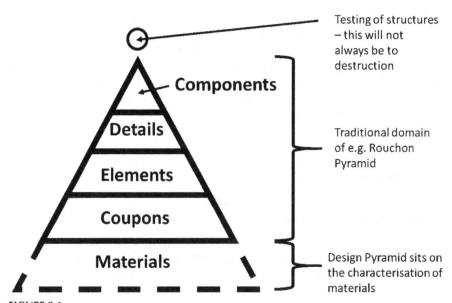

FIGURE 2.1

The Pyramid of Testing: An understanding of the behaviour of a structure or complex assemblage, such as an airplane, is built on a fundamental understanding of the behaviour of materials. This understanding helps with the design of coupons, and once these are tested, the information gathered informs the building of structural elements and more complex components.

the fibre architecture.[9] Rouchon (1990) moves on from these simple coupons to 'elements' (shapes produced with specific fibre architectures), through to 'details' (i.e., these shapes, but now with holes, changes in fibre architecture, or changes in thickness, the focus on this last being the effect of ply-drops). From details we move to sub-components (a box section or another part with structural and functional features) and eventually to components, e.g., an entire wing.

Rouchon's (1990) pyramid is exactly what is required in the context in which it is used, but understanding how a manufactured component will behave is dependent on a thorough understanding of the constituent materials. Composites have properties that change as a result of being combined – and so the first level of the pyramid should consider the constituent resins and fibres (although information on these is usually provided by the manufacturer) and leading to their combination. Different processing routes should also be considered at this stage, and the properties achievable that are dependent on how the material is manufactured. From coupons we move

---

[9] I.e., the directions of fibres with reference to principal stresses, the nature of the weave, if woven, or other method of keeping the fibres together so that they can be handled, and a component manufactured.

to elements. Again, consideration of processing is necessary: some combinations that are easy to produce as flat panels become problematic when we move to shapes. The addition of details can present further problems: will these be machined in after manufacture, or built in as part of the as-made piece? Machining composites is not easy, and can compromise the integrity of the piece, but adding in holes during manufacture, whilst achievable to a high tolerance, is still not straight forward.

2.5 The Role of Mechanical Testing: Capabilities and Limitations

This then brings us to the role of mechanical testing – and the question we started the chapter with: why break things? Materials are selected for various applications based on their performance in various categories. Not the least of these is how much the material, and any processing to get it into a particular form, will cost. Characterisation may therefore consist of understanding what is happening at the surface of the material, because this may affect how the material behaves under various environmental conditions, or how well it takes a coat of paint, or is bonded by an adhesive.

Mechanical testing is the suite of characterisation tools that allow us to determine the physical properties associated with applied loads. In engineering structures 'live' and 'dead' loads might be considered, which is to say loading that moves and varies, imposed by an operational condition[10] and the self-load of the structure; in the case of a water main, the dead load is the load of the pipe and the live load is anything associated not only with the water passing through, but with the soil cover, vehicles passing over the buried asset, ground frost, and many other factors.

There is a need to consider the durability of a material. Again, there are various things that might affect this, including damage occurring from radiation, such as solar ultraviolet light: there is a corollary between the types of damage that can develop over time in human skin tissue and that observed in most polymers, leading to a breakdown in performance over time.

So, we break things to understand how and why they break, to test their limits and impose safety factors. We look to see how materials and structures are damaged by their interaction with the environment, and what impact this has on their ability to fulfil their role, and how their capacity changes over time. In later chapters we will explore the various mechanical properties that

---

[10] Such us wind loading, as in the case of the Tay Bridge discussed in Chapter 1, traffic, the flow of fluids of granulated goods through piping, expansion, and contraction due to heating, and so on.

are likely to be of interest when testing a new material or determining the impact of in-service use against baseline performance.

References

Diamond, J.M., 1998. *Guns, germs, and steel: A short history of everybody for the last 13,000 years.* Random House.

MacKenzie, D., Zoepf, S. and Heywood, J., 2014. Determinants of US passenger car weight. *International Journal of Vehicle Design*, 65(1), pp.73–93.

OECD, 2010. *Materials case study 1: Critical metals and mobile devices.* OECD Environmental Directorate.

Richter, D. and Krbetschek. M., 2015. The age of the Lower Paleolithic occupation at Schöningen. *Journal of Human Evolution*, 89, pp.46–56.

Rouchon, J., 1990. Certification of large airplane composite structures. In *ICAS Congress Proceedings* (Vol. 2, pp. 1439–1447). International Council of the Aeronautical Sciences.

UNEP, 2009. *Sustainable innovation and technology transfer industrial sector studies: Recycling — From E-Waste to resources.*

3

Materials Science 101

Inventing is a combination of brains and materials.
The more brains you use, the less material you need.

– Charles Kettering

3.1 Introduction

There are innumerable textbooks on the subjects of Materials Science and Engineering, as well as on specific aspects both in respect of materials and on the techniques used by the practitioner to analyse them in various ways. Some of these books are incredibly niche, whilst others take a broader view of a particular topic. The purpose of this book is not to contribute to this myriad with yet another introduction to the intricacies of Materials Science, but it is important to understand some basic principles and how these affect the performance of particular materials, and consequently the performance of an engineering structure.

Take a group of light bulbs for instance, or a collection of water pipes, or pretty much anything that is manufactured in a batch. The group is nominally identical, but some will fail very early on, the majority will perform well for a certain amount of time, and a handful will continue to function for a surprisingly long time after the rest of the cohort have failed. Some will last so long, in fact, that they are taken out of service before failure, because it is cheaper to replace a cohort in one go, rather than piece-meal. Why is it that one item fails early, others last seemingly forever, and others fail at a point in-between these two extremes? Can we finesse things to move the failure curve? What causes these failures to occur in the first place?

Whilst it is possible to undertake a campaign of mechanical characterisation without understanding anything about the materials that form the structures being tested, it helps a lot to know something about the way that atoms work together to make the materials we use.

At its simplest, Materials Science can be described as the study of the behaviour of materials and how this behaviour is affected by the processing route. Consider Figure 3.1, this is a visual representation of this definition,

DOI: 10.1201/9780367822347-3

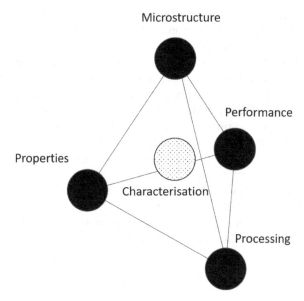

FIGURE 3.1
The Materials Tetrahedron, demonstrating the relationship between microstructure, properties, performance, processing, and the characterisation of materials.

sometimes referred to as the Materials Tetrahedron. In some senses, the figure is clearer, because it allows us to see all the factors in the relationship, and how they relate to each other, an embodiment of the maxim that a picture is worth a thousand words. The nodes of the tetrahedron are microstructure, properties, performance, and processing. Each of these areas of interest affects, and is affected by, the other three. In the centre of the tetrahedron sits characterisation. Such characterisation could refer to observation of the microstructure by optical or electron microscopy, a suite of spectroscopic tools, or the current focus, mechanical characterisation.

Why should this be of interest to the general engineer? Surely a material is a material is a material? Characterisation is of interest only to the specialist – once it has been done, that is the end of the matter…?

Let us consider an example. It would be difficult to estimate the number of different steels that are available. Total global annual production hit one billion tonnes early in the 21st century, and whilst there have been fluctuations, this has risen steadily, reaching two billion tonnes per annum by around 2020.[1] At

---

[1] The impact of the COVID-19 pandemic is still to be completely accounted for, but history suggests that any downturn in production will be a blip in the long term. In 2024, it was announced that a major UK steelworks would shut down production of primary steel (i.e. using a blast furnace to reduce mined ore and produce steel and other byproducts), causing significant national angst: what price a country that cannot produce its own steel? More broadly, and looking to the longer term, the production of steel is a major producer of CO_2,

its simplest we can consider the carbon content: steels are, fundamentally, an alloy of iron and carbon. The carbon content can be as low as 0.002 wt.% and as high as 2.14 wt.% – more than this and we move out of the range of steels and into the realm of cast irons. By varying alloying additions, processing routes, and heat treatments, it is possible to produce mild steels, stainless steels, and maraging steels, to name but three *families*. Most people would recognise the very different performance in terms of corrosion that would be exhibited by mild and stainless steels. What might be harder is determining which of the fifty-seven types of stainless steel is best for a particular application, not to mention ensuring that the microstructure is not compromised when the material is formed into the structure or component.

In the following sections, we will consider the nature of materials, beginning with a brief overview of the structure of the atom, how atoms bond together and how, as we increase the numbers of atoms collected, we see the creation of defects that will impact on performance.

3.2 The Atom

The concept of matter being composed of four (or sometimes five or even six)[2] entities, present in all things in different proportions, is one that crops up in ancient cultures around the world. This concept has taken various evolutionary routes, and observers have commented on the links between these ancient elements and the constituents of atoms.[3]

To the disgust of many Engineering students trying to understand the behaviour of a material, even in terms of mechanical characterisation, a knowledge of electrons, protons, and neutrons is essential. Much to their relief, we do not need to stray into the realms of particle physics and worry about the flavours of quarks or whether a hadron has a half-spin that is up or down.

However, things are not as bad as they might appear. The fundamental concepts that need to be considered are simply explained, and books are recommended for further explanation, if required, at the end of this chapter.

---

and this will need to be addressed if Net Zero targets are to be met. Do we need so much primary steel? Some will say yes, of course we do if we are to grow the economy, but the answer must be more nuanced if we wish to grow the economy sustainably. Better recycling practices, changes in the way that we use resources, and other factors could reduce the demand for primary steel, although the full discussion of this would take a book.

[2] The usual four are earth, air, fire, and water. Aether is sometimes added to the mix, and some cultures replace one or more items on the list, or add to it, with wood and/or metal.

[3] This is sometimes thought of in terms of the things such as electrons, protons, and neutrons, and sometimes in terms of the quarks and other fundamental building blocks of matter.

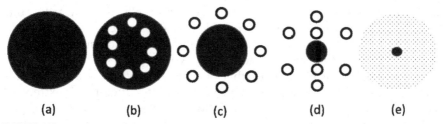

(a)   **(b)**   **(c)**   **(d)**   **(e)**

FIGURE 3.2
The evolution of human understanding of the atom. (a) Dalton's ball-bearing model (the atom is a discrete body with no internal characteristics); (b) JJ Thompson's plum pudding model (the atom is a discrete body with no overall charge, but regions within the atom carry negative charge, whilst the rest of the atom carries a counter-balancing positive charge); (c) Rutherford's model (negatively charged electrons orbit a positively charged nucleus; (d) Bohr's model, demonstrating that electrons can occupy different energy states; and (e) Schrodinger's model which shows that the 'orbit' of an electron is actually a volume in which there is a probability of finding it.

The modern story of the atom begins around 1803 with John Dalton's work on elements, and he proposed what is, today, commonly called the ball-bearing model: an atom is like a ball-bearing, indivisible, and atoms of different elements are ball-bearings of different sizes, densities and so on. We can mix them all up and separate them out again and they do not really change. This model is perfectly adequate, and gives us something with which we can work. In some circumstances though, it is completely wrong and unhelpful – it does not consider electrons, protons, and neutrons, for example. It took slightly over one hundred years from Dalton's work for these building blocks to be recognised and understood, and then things changed rapidly. Between 1904 and 1926, JJ Thompson gave us the plum-pudding model (the atom is still solid, but composed of volumes with positive and negative charges, like fruit dispersed in a cake), Rutherford showed that the electrons were separate and orbited a dense core, Bohr showed that the electrons had different energies, and that electrons could be shielded from the positive core, and finally Schrödinger showed that the electrons might be anywhere, although it is possible to define a volume where they are likely to be.[4] This last model is still considered to be the most accurate but, in many circumstances, it is preferable to stick with Dalton's or Bohr's models (Figure 3.2).

A great deal of research is still conducted on the fundamental behaviour of protons, neutrons, and electrons, especially the interactions between protons and neutrons, and the implications of the wave-particle duality of electrons, including the presence of electrons in the nucleus of an atom.

Happily, for the current purpose we can proceed with a relatively simplistic structure of the atom: the nucleus, the cluster of positively charged protons

---

[4] It should be noted though, that the neutron was not discovered until 1932 (Chadwick, 1932).

and neutral neutrons at the centre of the atom pulls electrons towards it but because electrons can only come so close to each other, electrons are only able to fill certain stable energy shells, called orbitals. The closest orbital to the nucleus is the 1s orbital and can hold up to two electrons.

The simplest atom is therefore that of hydrogen, which is a single electron orbiting a single proton;[5] as a consequence of its structure it is extremely reactive, seeking to bond with other atoms and so become more stable. Helium, the first of the 'noble' gases, is two protons, two neutrons, and two electrons; therefore, its outer shell is complete, and atoms of helium are almost completely unreactive because they are stable. As more protons and electrons become involved, the orbitals encompass greater volumes, and more electrons can be contained in each orbital. Energetic events can lead to electrons being ejected from the atom. In terms of particle physics, this is as far as we need to go, although it is worth noting that the energy required to remove an electron can be measured and the element present inferred from this characteristic. Further, the energy required to remove a particular electron is affected by its relationship to the atoms around it, and this can be used as an investigative tool, see e.g., Watts and Wolstenholme (2019). Beyond this, the nuances of electron orbitals can be ignored, except where they relate to atomic bonding.

3.3 Atomic Bonding

All primary bonding is about the sharing of electrons, and all secondary bonding is about the current owner of an electron holding on to it, but letting other atoms see it. Hence, there are three forms of primary bonds (metallic, ionic, and covalent),[6] whereas secondary bonding is an order of magnitude

---

[5] Deuterium and tritium are rare but naturally occurring forms of hydrogen which include one and two neutrons, respectively. Tritium is radioactive. This becomes important when we start to consider the impact of the operational environment on future performance, especially where such atoms are present in large quantities over time. For example, whilst most of the difficulties with electricity generated from nuclear fusion are associated with the physical reactions required, generating the initial starting conditions, and preventing breakaway reactions, considerable effort is also required in developing the materials that contain and control these reactions, and extract energy from them. Hydrogen embrittlement is a well-understood problem in many sectors, but there is still some way to go to be sure that there will not be unexpected issues with deuterium- and tritium-doped hydrogen causing the embrittlement.

[6] Which can be visualised as 'hand-grasps': if covalent bond are a firm hand-clasp, metallic bonds could be considered as a grip further up the arm, whilst ionic bonding is, in a sense, more like a 'monkey-grip' with only the fingers curled around, strong enough under some circumstances but easier to break in others. The analogy breaks down of course when we think of ionic solutions, where the ionically bonded atoms dissolve and are dispersed, albeit that they have claimed the electron or the hole, as preferred.

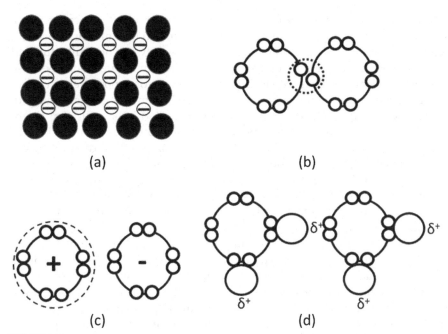

FIGURE 3.3
Primary and secondary bonding between atoms. (a) Metallic bonding, with electrons shared (and free to move) within the lattice formed; (b) covalent bonding, where two atoms complete an electron pair by sharing electrons; (c) ionic bonding, where one atom gives up its outermost electron and donates it to an atom with a hole in its orbital; and (d) an example of secondary bonding based on water, where a slight positive charge (δ^+) is put on the hydrogen atoms due to the covalent bond with oxygen. This δ^+ charge forms a weak bond with the electron pair of another oxygen atom.[7]

weaker than primary bonding (Figure 3.3): energetically speaking, secondary bonding is ~1/100th the strength of a primary bond, although exact bond strengths are dependent on a number of factors including the relative size of the atoms involved, relative polarity induced by the ease with which the electron is shared, and so on.

These forms of bonding affect most properties of materials, and their ultimate performance; the type of bond is a consequence of the behaviour of the electron around other atoms and electrons.

Metallic bonding: in metallic bonding, all the atoms that are collected together group themselves in nice, neat rows, and the electrons of the valence band, i.e., the most weakly bonded electrons, are free to move around amongst the collected atoms.

---

[7] More of a 'pinkie-finger promise'.

Ionic bonding: atoms with loosely bound outer electrons can enter a state that is more stable by allowing that electron to become disassociated. Similarly, some atoms have outer shells that are nearly, but not quite, full and so are willing to accept electrons and therefore become more stable. Because one loses a negatively charged electron, the resultant ion is positively charged, and conversely the one that accepts the electron becomes negatively charged.

In a low energy state, these charged atoms, called ions, will form themselves up in nice, neat rows like metals, leading to the formation of crystals. The most obvious example of this behaviour is table salt, sodium chloride, NaCl. This behaviour is not dependent on equal numbers of atoms, so long as charge parity is observed. So, for example, iron can form different oxides because it can lose two (Fe II/Fe^{2+}) or three (Fe III/Fe^{3+}) electrons. Oxides include FeO, FeO_2, Fe_2O_3, and there are also forms that mix Fe II and Fe III ions.

Covalent bonding: covalent bonding sees a sharing of electrons between two atoms. The difference compared with ionic bonding is that the orbitals overlap so that electrons are shared equitably. The usual example of this is carbon, which happens to have an outer shell which could hold eight electrons, but actually has four, and hence four spaces. It could give up those four electrons, or it could accept four to fill the gaps, but there is no strong driver to do either. Instead, an electron pair is shared between two atoms. Two carbon atoms could, therefore, form up to four covalent bonds. However, the greater the number of bonds, the greater the bond strain becomes, as the electrons are all pulled around to one side. Imagine that you are standing side-by-side with a friend, holding their hand. Now, without moving your feet or body, bring the other arm around to hold their other hand. It is possible, but places significant strain on your back and the arm reaching across your body.[8]

Secondary bonding, also called dipole bonding, arises from the movement of electrons. In an incomplete shell, a dipole can be created when an electron is shielded by the nucleus. Consider the Moon orbiting the Earth. The Moon cannot be seen from all points at once, and if we were orbiting the Earth at a point diametrically opposed to the Moon, we could not see it because the

---

[8] In the case of water, this secondary bonding has an important part to play in the states that water adopts, dependent on the overall energy in the system. Between 0°C and 100°C, the secondary bond between water molecules undergoes a process of formation and reformation which allows the molecules in a volume to move around relative to each other, but also gives the volume some coherence. Above 100°C and the energy in the system becomes such that secondary bonds are too weak to keep water molecules in contact with each other and the water becomes gaseous. Below 0°C and the energy in the system is so low that the water becomes solid. During the formation of ice, the secondary bonds take on a more permanent characteristic: the bond becomes more uniform throughout, so that the water molecules are locked in place in a regular pattern. It is a direct consequence of the secondary bonding between water molecules that leads to ice floating on water, rather than sinking, as would happen if the solidified water were more densely packed.

Earth would be in the way. In the case of an atom, with its central positive and orbiting negative charges, this means that there are times when there is a greater concentration of negative charge, and of positive charge, which can be 'seen' by other charged objects. In a similar manner to a magnet, opposites attract. In a covalently bonded molecule, there are regions where there is a permanent dipole, because the sharing of an electron pair means that electrons are held in position.[9] The usual example at this point is to consider water, H_2O. Oxygen can form two covalent bonds, hydrogen, one. A molecule of water is therefore made up three atoms forming two covalent bonds, giving rise to four permanent dipoles (Figure 3.3a), two negative (on the other side of the oxygen atom from the bonds, arising from already complete electron pairs), and two positive (one on each hydrogen atom). Hence, the three states of water (solid ice, liquid water, and gaseous steam) arise from the secondary bonding present, controlled by the level of energy in the system. When there is sufficient energy, the molecules vibrate so rapidly that there is little opportunity for dipole–dipole interactions to occur, and the water molecules move as far away as possible from each other, to fill the entire volume available to them. By contrast, when there is very little energy in the system, the molecules are unable to vibrate and move about, and the water molecules organise so that the negative dipoles on the side of the oxygen away from the hydrogen atoms attract the positive dipoles formed by the unbalanced hydrogen atoms of another water molecule (Figure 3.3b)

3.4 Crystal Structure

Not all materials form crystals, but crystal structure is an important part of many materials' transition from independent atoms and molecules to cooperative structures.

Recalling Figure 3.3a, some atoms prefer to be arranged in ordered arrays. If we take a cube, called a unit cell of these arranged atoms, at the simplest we would observe an atom at each corner (Figure 3.4a), which is referred to as a simple (or sometimes primitive) cubic unit cell. Each corner represents an intersection of eight unit cells; hence, each atom at each corner is shared between these eight unit cells, and hence one unit cell = one atom. The next kind of unit cell is the body centred cubic, where a single atom is present at

---

[9] The careful reader will have noted that this does not necessarily match with the previous comments about the probabilities associated with where an electron is located. The even more careful reader will realise that the level of complexity that this gives rise to is unnecessary in the current context. Whilst the briefest of summaries, the arguments laid out above are the convention for most basic chemistry and physics teaching and all that is required for this book.

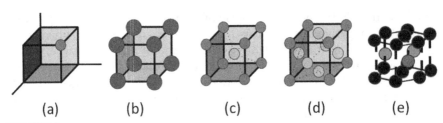

FIGURE 3.4

Examples of unit cells: (a) a unit cell is usually defined in an xyz cartographic system, where each corner is a node which is a junction between eight unit cells; (b) the simplest arrangement is the primitive cubic cell where there is only an atom at each node, which is shared between eight unit cells – the overall number of atoms associated with each cell is 1; (c) the body-centred cubic cell adds an atom in the centre of the cell – with the one atom associated with the eight nodes, the atom in the centre means that there is a total of two atoms associated with each unit cell; (d) rather than an atom in the centre of the cell, the face-centred cubic structure has an atom in each face of the cell, i.e. each atom (not at a node) is shared between two cells, and hence there are four atoms associated with each cell; and (e) the one exception to the cubic structure is the hexagonal close-packed cell, where there are six atoms associated with the cell, with the three in the central layer being completely within the cell, the two in the centre of the top and bottom faces being shared between two cells, and the atoms at the corners being shared with six unit cells, hence there are six atoms per unit cell.

the centre of the unit cell, in addition to the eighths of an atom present at each corner. So a body-centred cubic cell has two atoms per cell. Instead of placing a single cell in the centre of the cell we can place half an atom in the centre of each face, giving a face-centred cubic cell, with four atoms associated with each cell.

There is one further kind of until cell, which is hexagonal rather than cubic: hexagonal close-packed structures.

The full complexities of crystal structure and the terminology associated with describing lattices, which occupies entire textbooks and is the subject of entire courses within Materials Science degrees, cannot be summed up in a few paragraphs. However, the basics presented here are sufficient to allow us to understand that:

1. Atoms can be arranged in lattices.
2. Different lattice structures are possible with different configurations of atoms[10].
3. The different configurations mean that the atoms interact with different numbers of atoms.

This gives us the grounding we need to explore the issues surrounding defects in crystal structures.

---

[10] One of the complexities that we have not explored here is that these lattice structures are also relevant to alloys, i.e. mixtures of different kinds of atoms, and molecules.

3.5 Defects in Crystal Structures

In terms of understanding failure of materials, knowledge of the 'errors' that can occur in the arrays of atoms is critical. However, it is also important to remember that whilst such errors have an impact on materials properties and performance, the impacts can be positive as well as negative, and a part of materials engineering is to control the introduction and distribution of such errors. In this context there are four fundamental errors that we are considering (Figure 3.5a–d):

a. a vacancy in the lattice;
b. an interstitial insertion in the lattice;
c. a substitution in the lattice by an impurity; and
d. an interstitial insertion of an impurity in the lattice.

The overall effect of the introduction of an impurity will be dependent, to some extent, on the nature of the impurity (in particular relative size, which will affect the overall volume[11] affected; relative electron activity is also of interest as disparity can lead to localised disruption in electrical[12] fields), and the number of impurities present.

This is a very simple treatment of defects in crystal structures and, as noted on other aspects, there are textbooks on the subject of defects in crystals, some of which focus on only dislocations (e.g., Hull and Bacon, 2011), which can become more complicated when considered in three dimensions, or where the dislocation breaks the surface rather than simply being in the bulk of the material.

3.6 Micro, Meso, Macro: The Different Levels of Structure

This book considers structures as things that are made from materials, but in the context of materials themselves, there are different levels of structures that can control the properties of materials. At the macroscale, a component has features that are clearly visible to the naked eye, whilst at the

---

[11] The tendency is to visualise basic crystal structures in two dimensions, as this makes it easier to see the point of interest, but of course bulk structures will have multiple layers within the crystal, which we begin to see in Figure 3.3.
[12] And in some contexts, magnetic fields.

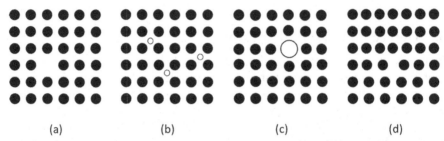

FIGURE 3.5

Examples of defects observed in crystal structures: (a) a vacancy in the lattice; (b) an interstitial insertion in the lattice; (c) a substitution in the lattice by an impurity; and (d) an interstitial insertion of an impurity in the lattice.

microscale, the features of interest cannot be observed without the use of magnification.[13]

This definition may appear slightly insipid – it is not based on physical measurements, for example – but there is a great danger in pinning things down too much. It has been observed that in the case of nanomaterials, for example, there is some leeway in deciding whether a phase is nanoscale or sub-micron (Jesson and Watts, 2012). An arbitrary definition, popular with physicists, is that nanoscale is anything that is below 100 nm in length; however, it would seem to be more important to base a definition on the effect that it has on properties. On that basis then, a nanoscale phase is one which is dominated by surface rather than bulk properties. For the current purposes however, nanoscale phases can be subsumed into a consideration of microstructure.

The mesostructure of a material is frequently overlooked or ignored, being subsumed into consideration of the micro- or macrostructure, as appropriate. However, the terminology comes into its own in certain circumstances, such as the incorporation of designed porosity, where the pore size is small, but the control of the size and dispersion is critical for the component being manufactured, e.g., for optoelectronics (Hayward et al., 2001) or biomedical applications (Xu et al., 2009). The concept of the mesostructure is also important in the context of composite materials, where two or more phases interact and can therefore lead to a range of defects (Yurgartis, 1995).

With the macrostructure, we are on more familiar, observable territory: we are dealing with something that is close to a complete component and we can begin to ignore the constituent atoms and molecules and phases of material.

---

[13] Some definitions of microscale, or microstructure, suggest that features within the microstructure require a minimum of 25× magnification to be observed. This is to say that if we were to observe a human hair, of the order of 100 μm in diameter, under a microscope, then anything that is observable only when the hair is magnified to 2500 μm (or 2.5 mm) in diameter counts as microstructure.

However, the gross defects that we can observe at the macroscale can lead to microstructural defects: large cracks can become atomically sharp and set the stage for catastrophic fracture.

A familiar illustration of this hierarchy of structure is the tree, and the useful timber that can be extracted. The whole tree is the macrostructure, and we can extract parts as necessary for industrial use. We can observe the grain, and there will be variation based on the type of tree felled, its age, location, and the history of the environment during the time it was growing. At the microstructural scale we can look at individual plant cells and the way that molecules[14] interact to form different parts of the cell wall. However, in between there is a rich world of detail, incorporating the formation of annual growth rings, the living sap wood, the heart wood, which is essentially a repository of unwanted chemicals, radial cells, and so on. Depending on what you are looking for, the mesostructure can be the most interesting part to look at.

3.7 Summary

Materials and structures fail for a variety of reasons, and in the event of a disaster, there are likely to be a number of events that coincide: poor materials selection, poor condition monitoring, and human error can turn a mistake in production into a significant loss of life. However, by understanding materials behaviour, by making good choices, by testing materials beyond their normal performance, we can mitigate, to some extent, human error. Beyond the incidence of catastrophic failures, by understanding the nature of materials, we can make an assessment as to the feasibility of extending operational lifetimes. By looking at the environmental conditions seen during the service lifetime, by reviewing the loading we might expect to see we can predict the future performance of the asset when it approaches the end of its design life, or, as is increasingly common, the asset is exposed to forces that were predicted to occur only once in a hundred or a thousand years. Key to this is understanding how materials behave.

This chapter has not considered the nuances of electrochemistry, for example, and the impact that galvanic corrosion can have on the lifetime of a component when poor material choices are made. This chapter has not considered the impact of magnetic or electrical fields. When getting into the detail of operational life these are certainly factors that need to be considered,

---

[14] Technically macromolecules, because as molecules go, they are very long. This is one of those situations in science where a linguistic toolkit based on certain well-understood principles can become confusing when different contexts at different length scales begin to overlap. If we are operating at the macroscale, we cannot observe macromolecules.

amongst others. But fundamental to the understanding of materials performance is bond strength, and how this can vary, crystal structure, and the impact that this has on the ability of the material to deform, or not, and the relationship between performance, property, processing, and microstructure (recall Figure 3.1).

References

Chadwick, J., 1932. Possible existence of a neutron. *Nature*, 129(3252), pp.312–312.

Cottrell, A., 1997. *An introduction to metallurgy*, Second Edition. Edward Arnold.

Hayward, R.C., Alberius-Henning, P., Chmelka, B.F. and Stucky, G.D., 2001. The current role of mesostructures in composite materials and device fabrication. *Microporous and Mesoporous Materials*, 44, pp.619–624.

Hull, D. and Bacon, D.J., 2011. *Introduction to dislocations*, Fifth Edition. Elsevier.

Jesson, D.A. and Watts, J.F., 2012. The interface and interphase in polymer matrix composites: effect on mechanical properties and methods for identification. *Polymer Reviews*, 52(3), pp.321–354.

Watts, J.F. and Wolstenholme, J., 2019. *An introduction to surface analysis by XPS and AES*, Second Edition, John Wiley & Sons.

Xu, F.J., Neoh, K.G. and Kang, E.T., 2009. Bioactive surfaces and biomaterials via atom transfer radical polymerization. *Progress in Polymer Science*, 34(8), pp.719–761.

Yurgartis, S.W., 1995. Techniques for the quantification of composite mesostructure. *Composites Science and Technology*, 53(2), pp.145–154.

Recommended Reading

Now in its fourth edition, Ashby and Jones is indispensable for anyone looking to ground themselves in the basics of materials and materials behaviour. Part 1 address properties applications and design, whilst part 2 considers microstructure and processing.

Ashby, M.F. and Jones, D.R., 2012. *Engineering materials 1: An introduction to properties, applications and design*, 4th Ed., Butterworth-Heinemann (Elsevier).

Jones, D.R. and Ashby, M.F., 2012. *Engineering materials 2: An introduction to microstructures and processing*, 4th Ed., Butterworth-Heinemann (Elsevier).

Another excellent undergraduate textbook on Materials Science and Engineering is Callister; modern editions come with useful digital tools, but even an older edition will stand the reader in good stead.

Callister Jr, W.D. and Rethwisch, D.G., 2020. *Callister's materials science and engineering*. 10th Ed. John Wiley & Sons.

For those who feel that they need to expand their knowledge of physical chemistry in support of the basics presented in this chapter, *Atkins' Physical Chemistry* is a standard textbook on many courses, although the lighter *Elements of Physical Chemistry* may be preferable.

Atkins, P. and De Paula, J., 2013. *Elements of physical chemistry*. Oxford University Press, USA.

Atkins, P.W., De Paula, J. and Keeler, J., 2023. *Atkins' physical chemistry*. Oxford University Press.

In a footnote, I raised the issues of the continuing trajectory of the growth of the steel industry. If there is a desire to pursue this line of thinking further, then two texts that are worth looking at are Julian Allwood's *Sustainable Materials* and Tim Jackson's *Prosperity without growth*.

Allwood, J.M., Cullen, J.M., Carruth, M.A., Cooper, D.R., McBrien, M., Milford, R.L., Moynihan, M.C. and Patel, A.C., 2012. *Sustainable materials: With both eyes open.* Cambridge, UK: UIT Cambridge Limited.[15]

Jackson, T., 2016. *Prosperity without growth: Foundations for the economy of tomorrow.* Taylor & Francis.[16]

---

[15] Republished in 2015 by Bloomsbury as *Sustainable materials without the hot air: making buildings, vehicles and products efficiently and with less new material.*

[16] This is the second edition, revised and updated from *Prosperity without growth: Economics for a finite planet* published by Routledge in 2009.

4

Defects and Degradation

"One sees qualities at a distance and defects close to."

– Victor Hugo

4.1 Introduction

In Chapter 3, the concept of defects within crystal structures was introduced. These are defects writ small, at the level of the misplacement of atoms, but which in quantity and repetition can have a significant effect on the behaviour of the material and hence the structures into which they are made. The concept of different levels of structure was also introduced, with this interest in the placement of atoms being at the microstructural (or occasionally the nano-structural level). In this chapter we will explore the defects that are found at the macrostructural level. These defects can be split into three categories:

1. Defects produced during manufacture.
2. Defects induced during installation.
3. Defects arising throughout service.

But what then is degradation, and what is the difference between this and a defect? In the context of the last point on the list above, we could use degradation as a special case, although defects that arise throughout service can be temporally discrete (such as an impact leading to plastic deformation or cracking), or temporally continuous, such as the cracking engendered through the freeze–thaw cycle of water entering a non-critical defect, becoming ice and causing the defect to grow: this process may occur hundreds or even thousands of times over the course of years, decades, or centuries before the defect becomes a critical one.

DOI: 10.1201/9780367822347-4

The issue is exacerbated by the fact that whilst 'defect' can only be a noun, degradation can be both a noun and a verb. In the context of this book, i.e., the failure of materials and structures we can use the following definitions:

Defect, noun, a feature of the structure of a material that is different from the general configuration, and which will give rise to localised mechanical properties that are different from the bulk. Defects may be critical or sub-critical, where a critical defect is defined as one which will cause failure if the loading on the part or structure reaches the yield or ultimate strength of the material in the location of the defect. A sub-critical defect is one which is small enough that it will not cause failure during normal operation, but which has the capacity to grow over time, e.g., because of fatigue, and become critical at some point in the future.

Degradation, verb, any of a number of processes including but not limited to corrosion and wear, which will eventually lead to the failure[1] of the material, component, or structure. The effects of degradation can be observed, in some circumstances, before failure occurs, although it is not usually possible to quantify these effects by eye.[2] Specialist equipment[3] will need to be used to, e.g., measure the depth of degradation, or its impact on electrical resistance or conductivity, for example.

Degradation, noun, the impact of a process of degradation on a component or structure.

Or, to put it another way, not all defects are the result of degradation, and not all degradation leads to a defect that we would see as a problem in the context of, e.g., fracture mechanics (see Chapter 6). There is, however, certainly some overlap (Figure 4.1).

The first step to understanding the impact of a defect is to understand how it comes about. This has been explored in Chapter 3, at least in as much as it relates to the microstructure. The next section considers macro-scale defects and their origin, and afterwards some examples are presented of how to eliminate defects, or how to live with them.

---

[1] Recalling that failure need not be catastrophic, but simply that a part is no longer able to perform its function, or even that it no longer meets an operational standard. For example, if the standard calls for an unbroken layer of paint on the surface, and the paint becomes scratched or worn through, it has failed, even if the component it is on is still able to function normally.

[2] Some may argue that you can look at a surface and determine that it is worn. This may be true, to an extent, but comparing two items from the same batch and determining which is the more worn is trickier and determining which (either? both? neither?) is critically worn when compared with a standard is trickiest of all. Even experienced practitioners struggle to rank defects across a surface. And this is just dealing with surface defects and degradation – what about those on hidden surfaces, or those within the bulk volume of the component of interest?

[3] Occasionally, specialist equipment can actually be very basic, although it tends to be more destructive in the long run: if you can break a piece off something with either your bare hands, or brute force applied through a chisel or even a broom handle, you can get a feel for how badly degraded something is. If you are at that point though, it is probably too late.

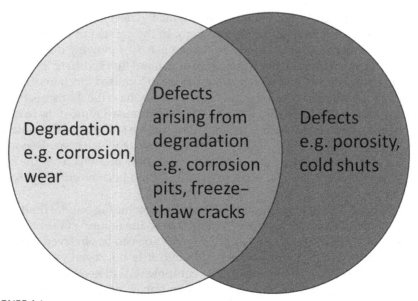

FIGURE 4.1
Venn diagram demonstrating overlap of defects and degradation: the terms are connected but are not synonyms.

4.2 An Unofficial Taxonomy of Macro-defects

4.2.1 Overview

For illustrative purposes, the next section considers a cast iron component, specifically a section of pipe for a water main, and the defects that can affect its performance. Cast iron's hey-day may have been the Victorian era, but it is still used today in a variety of applications, not to mention legacy assets that must be maintained. Its inherent microstructure (the formation of graphite flakes, see ASTM A247-19) and the manufacturing processes give rise to defects that can be considered representative for the purposes of this chapter.

4.2.2 Manufacturing Defects

The oldest cast iron was cast in a pit, whereby dross, dissolved gas, and the like would rise to the highest point. As the melt cooled, the dissolved gases could no longer remain in solution and so would form bubbles, which would rise through the melt until the melt was too cool for the bubbles to be able to move, leading to porosity. This often meant that a line of weakness could be introduced if the component were cast horizontally; later, castings were placed vertically and a sacrificial portion which contained all the potential defects could be removed.

Prior to the introduction of materials standards, and indeed prior to a full understanding of the nature of iron, cast iron, and steel, and the introduction of mass production, it was not unusual for small foundries to pop up where they were needed, move on when they were not, and for production methods to be controlled by instructions passed on by word of mouth. Hence, control of temperature throughout the pour and production of the components could be variable, and the additions to the melt to control various aspects would be done by eye, or judgement.[4]

The properties of cast iron are controlled by the graphite flake structure (Angus, 1976), and so it is not surprising that variable properties were observed even after taking the variable defect population into account.

Manufacturing defects are sometimes referred to as *ab initio* defects, literally 'from the beginning'. Hence, *ab initio* defects are those 'baked in' during manufacture; these include, but are not limited to:

- microstructural defects;[5]
- porosity;
- slag inclusions (from the source material);
- impurities and contamination from the manufacturing site and improperly maintained equipment;
- cold shuts (where two fronts of molten iron fail to fuse properly due to premature cooling); and
- geometrical issues arising from improper alignment of moulds,

Whilst we have been using cast iron as an example, it can be understood that these issues can arise in any poorly conducted manufacturing endeavour, with any material which is not treated appropriately. In practice, these sorts of defects are much rarer, and the quality assurance processes are more robust, ensuring that items with such defects are now much more seldom encountered. However, where small items are produced in large quantities, such as screws and other fastenings, it can be interesting to look at items that fail when they are first used in an attempt to identify a defect that has led to failure.

---

[4] Recycling was surprisingly common too, and at one time it was not unusual to find unmelted bits and pieces of prior incarnations embedded in a casting, including unburned lumps of wood.

[5] Cast iron is probably the poster child for this issue. Carbon was added to iron to help reduce the temperature at which iron melts and hence make casting easier. However, the flake structure forms when the iron cools and the carbon is rejected from the melt. The flakes act as crack initiators, significantly affecting the tensile strength of cast iron.

4.2.3 Installation Defects

Following manufacture, cast iron pipes used in the water industry were transported to site for installation on trucks and trailers and would either be rolled down planks of wood from the edge of the truck or be lifted with a crane. At this point, the early quality assurance method would have been to strike the pipe with a hammer to cause it to ring.[6] Rajani and Kleiner (2010) suggest that pipes handled in this way were susceptible to damage from collisions between the pipes and other external bodies. Their argument is that not all such impacts would produce a visible crack: an internal crack in the bell or spigot would not necessarily be detected by on-site hammer testing. Whilst such damage may have occurred, and could be an issue in other contexts, their further contention that such cracks would grow under fatigue loading over time is less credible for three reasons:

1. The cracks would need to be a very precise length not to cause immediate failure, and to grow under fatigue cycling over the course of the time that we know such pipes have spent in service.

2. Cracks in the bell, and to some extent the spigot, are less of an issue than other defects.

3. The loads required to grow such cracks would have to be very precise – any less and nothing would happen, any more and other defects would be more of an issue.

Other defects arise from the use of chains for pipe handling rather than canvas slings: this is known to have damaged the protective outer coal tar or bitumen coating, leaving areas of bare metal, which would then be prone to localised corrosion.

Once the pipes were *in situ*, the joint would be packed with hemp to form a seal, filled with molten lead and finally caulked by hammering the solidified lead into the joint (Rajani and Abdel-Akher, 2013). If the caulking were carried out incorrectly, the lead would be hammered too far into the joint, leading to an increase in stress around the joint, potentially causing a crack to initiate. A worst-case scenario would see poor caulking practice expand a crack created during the siting of the pipe. This could result in a structurally flawed pipe with a reduced service life.

Generalising, we can see that there are a range of ways in which defects can be introduced after manufacture but before use. Many components are left stockpiled for a time, at the point of production, at storage facilities

---

[6] In much the same way as a bell. A similar method was used for checking the integrity of the wheels of steam locomotives in service, a fact that provided a vein of humour for observers and comedians in late 19th and early 20th centuries.

belonging to the manufacturer or the user, and potentially on site. Defects and proto-defects can therefore be induced by degradation arising from, for example, UV radiation from daylight, humidity, chemical species in the atmosphere, or animal attack. The list continues to grow as we continue to gain understanding of materials health and what might affect it. Different materials will be more affected by one mechanism or another, but fundamental considerations when examining a failure, or attempting to prevent a failure from occurring, are:

- Handling: is there a danger of the component being damaged when being manoeuvred from where it is manufactured to where it is moved?
- Environmental: in the widest sense, anything that is not associated with the handling of the component but arises from the properties of the atmosphere and the wider world.
- Installation (and 'making good'): anything arising because of the process of commissioning the component, particularly where two or more components are joined together. Defects can be created due to over-heating of a polymer seal for example, or could be created by the component being 'grazed' or dented as people work around the area.[7]

4.2.4 Service-induced Defects

Once in service, water mains are subject to further mechanisms that can cause defects to form and grow until the defects are too large for the pipe to sustain. In places where the bitumen or tar coating is compromised (often because of the installation process as mentioned earlier), corrosion processes are initiated which can further undercut the coating, causing it to spall, expose bare metal, and continue to corrode. Of particular concern is 'graphitisation', which should properly be termed 'graphitic corrosion' (Logan et al., 2014a, 2014b). Despite anecdotal assertion, it is not possible to identify graphitic corrosion by visual inspection in exposed pipes since the corroded region does not change in volume (although the density decreases) and the pipe appears unaffected to the naked eye.

Here, then, is where degradation really begins to come into its own: in engineering a component will be designed such that there is a safety factor

---

[7] An example from personal experience would be the carpenter fitting the handrail on some new stairs and drilling through a water pipe. A very visible and obvious failure, but it can be imagined that a glancing blow or damage from a drill that fails to penetrate the wall of the pipe could sit quietly for some time until an inciting incident caused a more problematic failure. Water escaping from such a pipe could easily cause more damage and degradation, leading to bigger problems with the building.

applied to the part, and the final design will encompass not only the general operational requirements, but the loads that may be placed on the component in a predictable emergency,[8] and some level of safety factor. However, the capacity for a structure to carry load can be degraded by biological, chemical, or mechanical action. In some instances, the loss of capacity can be calculated simply by determining a ratio of the thickness of the material affected with the original thickness, as a so-called loss-of-section approach. However, where the degradation leads to cracking or other penetrating damage such as corrosion pits, even very little damage can lead to catastrophic failure, as the damage leads to stress concentration. Jesson et al. (2013) demonstrate the different effects that the same basic corrosion mechanism can have on a material when the corrosion is generalised (loss-of-section) or more pit-like (fracture mechanics). As an aside, but in keeping with the concept that understanding failure can lead to better materials modelling, and therefore more focused mechanical test programs, Fahimi et al. (2016) and Ugoh et al. (2019) were able to use this understanding to provide predictive models.

Of course, life is never entirely straightforward. Having developed a nice, neat classification for the description of different kinds of defects, we immediately come to those that don't fit nicely into this system. For example, if we look at barely visible impact damage (BVID[9]), this is something that is usually associated with the operational phase. In some respects, it looks a lot like some of the issues discussed in the section on installation defects. BVID is usually characterised as being low energy, of the order of 15 J and below, and is often illustrated as being the situation where someone effecting repairs on a structure drops a tool which hits the surface, possibly, but not necessarily, leaving a mark or very small dent, but otherwise apparently no damage. The important damage lies below the surface. In the context of composite materials, the key issue is the cracking that can occur in the matrix. Whilst it begins by being sub-critical, this damage can grow by fatigue once the component returns to service. It is not really a service-induced defect as described at the beginning of this section, but neither is it an installation defect.

---

[8] Predictable in the sense that it is possible that it will occur, not necessarily that it is probable. For example, the chances of a car driving into a static object such as a telephone pole, or a building, are on the whole unlikely, but not, as we know from the news, impossible. Under normal circumstances, if the object has been built or sited correctly, the event should not occur, but factors such as the driver speeding, or a previous spill on the road, or an animal leaping out at an inopportune moment can lead to a crash.

[9] BVID is usually considered in the context of composite materials, but there is potential for it to occur in other contexts as well.

4.3 Microstructural Defects: How to Avoid Them or Turn Them to Your Advantage

4.3.1 Overview

Whilst it is possible to eliminate defects through better manufacturing controls, better handling of the components between the production line and installation (via storage), and optimised operational controls, microstructural defects remain. However, there are further opportunities to address defects, potentially eliminating them completely or by turning them to our advantage. Here we consider four such opportunities: single crystal structures, fibreisation, nanocrystal structures, and work hardening.

4.3.2 Single Crystals

Recalling from Chapter 3 that many materials, predominantly metals and their alloys but including a plethora of others, such as ceramics, and to a certain extent some polymers, form crystals, and that these crystals are prone to defects, what can be done about this? One approach would be to do nothing at all. We can characterise the defects, understand their behaviour, and live with their consequences. In many situations this is exactly what is done. However, consider, for example, a turbine blade, such as that found in a jet engine. Such a blade:

1. operates in a challenging environment;
2. is expensive; and
3. is safety critical.

If we can improve the performance of a blade, by reducing the defects that impact its mechanical performance, then we can increase safety and reduce the amount of downtime arising from maintenance. Whilst there is only so much we can do about the packing defects that can arise, during the formation of a crystal, one thing that we can do is to produce a component with no intergranular defects by eliminating grain variation by ensuring that the crystal is a single grain (Figure 4.2).

So, by manipulating the manufacturing process to ensure that only one grain is initiated as the melt cools, we can create a turbine blade that is made from a single crystal. The initiation of the single crystal involves the careful manipulation of the cooling melt using a 'pig's tail' shaped entrance to the mould, which prevents all except one grain developing further (Figure 4.3). This part is then cut off to leave a single crystal.

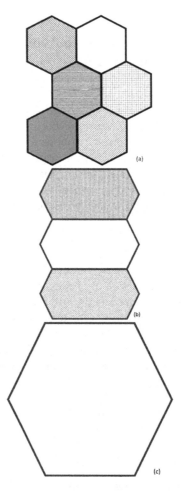

FIGURE 4.2
Idealised grain structures within a crystal: (a) the crystal is made up of a number of grains with various orientations arising from independent initiation of grains within the crystal; (b) grains are aligned in a specified direction although multiple grains with different orientations are still present, each grain runs the length of the component; and (c) the crystal is formed from a single grain derived by control of the initiation and/or elimination of competing orientations.

Counterintuitively, the removal of grain boundaries from the structure generally reduces the strength of the material, but it eliminates sites that could initiate crack growth, and it improves the creep performance of the blade, both significant wins in the overall performance of the blade and hence the engine.

Jet engine turbine blades are not the only use for single crystals, and there are currently eleven main routes to producing single crystals from various elements and compositions for electronic components, synthetic gemstones, and engineering parts. These methods fall into four main categories, depending

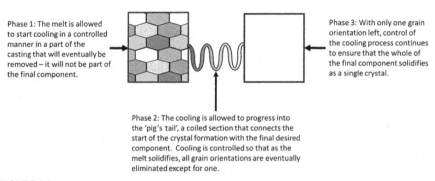

Phase 1: The melt is allowed to start cooling in a controlled manner in a part of the casting that will eventually be removed – it will not be part of the final component.

Phase 3: With only one grain orientation left, control of the cooling process continues to ensure that the whole of the final component solidifies as a single crystal.

Phase 2: The cooling is allowed to progress into the 'pig's tail', a coiled section that connects the start of the crystal formation with the final desired component. Cooling is controlled so that as the melt solidifies, all grain orientations are eventually eliminated except for one.

FIGURE 4.3
Illustrative example of the creation of a component formed from a single crystal.

on whether the crystal is grown from solid, melt, solution, or vapor, with different 'flavours' arising from processing temperatures and other variations in the overall process. Synthetic production of various mineral structures has been of interest since around 2500 BCE, beginning with the purification of salt,[10] continuing through the early understanding of materials and attempts to recreate what was found in nature and arriving, eventually, at the need for high-grade electronics and optical components. Today, the largest producer and consumer of high-grade, defect-free, large-scale single crystals is the semiconductor market.

4.3.3 Fibreisation

A fundamental aspect of the production of composite materials, at least those based on fibres, is that by turning a three-dimensional volume into a 2.5-dimensional one, critical defects can be eliminated. Consider an expanse of glass (Figure 4.4a): for the purposes of this example, we can ignore the thickness. The glass is placed into omni-directional stress, i.e., an equal amount of the tensile stress in every direction (Figure 4.4b). Based on our calculations, this should be sub-critical, yet the glass shatters. Perhaps an issue of the stress state? This being a thought experiment, we can use the same sheet of glass again. Let us make the stress-state simpler, still tensile, still apparently sub-critical, but only applied in a single direction. The glass breaks. We have nothing but time, and our thoughts, so we repeat this situation, testing the plate by applying tensile stress in different directions. Every time, the sheet of glass breaks at an unexpectedly low load (Figure 4.4c).

The problem is that the glass is full of defects that act as stress concentrators. These defects are randomly dispersed within the sheet of glass, vary in size,

---

[10] Recalling of course that salt was an essential preservative and flavour enhancer, and of course that the word salary derives from the daily salt ration that was given to Roman soldiers.

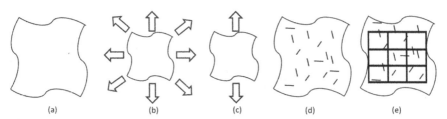

FIGURE 4.4

The mechanical properties of a piece of glass – why does it keep failing? (a) A typical piece of glass, (b) load the glass equally in all directions, (c) simplify the load case to a single tensile load, (d) we become aware of a population of microstructural defects preventing the glass from achieving its full potential, and (e) we reduce the impact of the defects by cutting the glass into smaller pieces, eliminating some defects completely and reducing the size below critical of others.

shape, orientation (Figure 4.4d), and can arise from various causes including dust, porosity, uneven cooling, and the like. However, if we can eliminate or at least mitigate the presence of the defects, we will have a stronger piece of glass. If we cut the sheet into smaller pieces, we can play a sort of Russian roulette, by reducing the chance of a critical defect – although we may introduce more defects in the cutting process (Figure 4.4e). Make the pieces smaller still and we stand a good chance of eliminating or isolating more critical defects. However, there is then the danger that we make the pieces so small that they are no longer useful.

By casting fibres from the melt, we create long, thin pieces of glass, which are too small to contain defects that are problematic should the glass fibre be placed in tension. By bundling these fibres together, we can create something that approaches the maximum strength of glass. For those interested in composite materials, we then need to consider the manner in which we align the fibres in order to deal with complex stress states – if our fibres are all aligned in a single direction, then the overall material will be strong parallel to the fibres but incredibly weak perpendicular, having only the strength of the resin (adhesive) that is used to bond the fibres together – the matrix in a fibre-reinforced polymer matrix composite.

So far, we have considered only glass fibres, and by extension some of the other mineral fibres derived in a similar way, such as basaltic fibres; however, a similar approach can be taken with carbon fibres, and those formed of polymers and other macromolecules. For example, spider silk is known to be incredibly strong, and we can therefore intuit that at least part of the performance of the silk comes down to the size of the fibre, typically 2.5–4 μm in diameter. Currently, commercial attempts at wet-spinning various synthetically produced proteins into silks can 'only' achieve diameters of around 10–60 μm. These are giving rise to a range of properties which are comparable with naturally produced spider silks, but it can be understood from the discussion above that any attempt to produce a sheet of material will give rise to the generation of defects that will impact on the performance of the material.

Armour, and other applications, remain an aspiration: once the question of mass production has been addressed, there remain many phases of testing, accreditation, and acceptance to be completed.

4.3.4 Work Hardening

Work hardening is a process that has been used for thousands of years, without it being understood why it works, or indeed how. If you have ever seen a blacksmith at work, you will know that it involves a lot of hitting the piece of metal being worked on. The beginning phase of producing something in this way is a lot like tenderising a piece of meat. In a piece of meat, especially a muscle that is used a lot, the underlying muscle fibres become dense and difficult to chew and digest. This fibre structure can be broken down by chemical action (marinading) or by blunt force, with a meat hammer, or similar. A 'raw' piece of metal is similar: whilst there is no fibre structure in the same way, there are impurities: the early smiths realised that these impurities could be beaten out of the metal. This is useful, because it means that you do not have to melt the ore completely to remove the impurities.

Once you have a piece of metal that is ready for use, it will then need to be shaped. This will require further application of a hammer, but now the bending and shaping of the metal will introduce dislocations, as discussed in the previous chapter. Some of this change in the microstructure will be useful, but it can also make the material more brittle. Again, recalling from Chapter 3 that there are different ways in which atoms can be oriented in relation to one another, the different crystal structures make it easier or harder for dislocations to move. Imagine a game of three-dimensional chess where different metals behave like different pieces in the game. Some pieces have more degrees of freedom, others less; some can move between levels and others cannot.

Aluminium is an example of a metal with few degrees of freedom, and hence work-hardening is to be avoided – it will embrittle the component leading to failure. Copper on the other hand can be deformed significantly before work hardening becomes a problem, and so its strength can be increased with minimal impact on toughness.

4.4 Summary

Defects can be problematic, but they can also have their advantages. Fundamentally though, by understanding that defects:

1. exist,
2. can be accounted for,

3. can be detected, and
4. can be eliminated

we can go a long way to understanding failure. However, when designing a component based on this knowledge, beware of unintended consequences. If you eliminate a known set of defects, what will the consequences be? Will you introduce a new set of unknown defects? Will you introduce a new failure mechanism?

The word defect carries with it a lot of negativity, and we can point to defects that have caused failures. However, a defect is, fundamentally, just a volume of a material that is not like the rest, and therefore causes a localised difference in properties. Whether this is a problem or not, might depend on the prevalence of this localised difference, or whether by knowing of its presence we can utilise it – or not.

References

Angus, H.T., 1976. *Cast iron: Physical and engineering properties.* Butterworth London, UK.

ASTM A247-19, 2019. *Standard test method for evaluating the microstructure of graphite in iron castings.* ASTM.

Fahimi, A., Evans, T.S., Farrow, J., Jesson, D.A., Mulheron, M.J. and Smith, P.A., 2016. On the residual strength of aging cast iron trunk mains: Physically-based models for asset failure. *Materials Science and Engineering: A*, 663, pp.204–212.

Jesson, D.A., Mohebbi, H., Farrow, J., Mulheron, M.J. and Smith, P.A., 2013. On the condition assessment of cast iron trunk main: The effect of microstructure and in-service graphitization on mechanical properties in flexure. *Materials Science and Engineering: A*, 576, pp.192–201.

Logan, R., Mulheron, M.J., Jesson, D.A., Smith, P.A., Evans, T.S., Clay-Michael, N. and Whiter, J.T., 2014a. Graphitic corrosion of a cast iron trunk main: implications for asset management. *WIT Transactions on The Built Environment*, 139.

Logan, R., Mulheron, M.J. and Jesson, D.A., 2014b. *Observations on the graphitic corrosion of cast iron trunk main: Mechanisms and implications,* Eurocorr 2014: European Corrosion Congress Pisa, Italy.

Rajani, B. and Abdel-Akher, A., 2013. Performance of cast-iron-pipe bell-spigot joints subjected to overburden pressure and ground movement. *Journal of Pipeline Systems Engineering and Practice*, 4(2), pp.98–114.

Rajani, B. and Kleiner, Y., 2010. Fatigue failure of large diameter cast iron mains. In *Water Distribution Systems Analysis*, pp.1146–1159. American Society of Civil Engineers.

Ugoh, G., Cunningham, R., Farrow, J., Mulheron, M.J. and Jesson, D.A., 2019. On the residual strength of ageing cast iron wastewater assets: Models for failure. *Materials Science and Engineering: A*, 768, pp.138221.

5

Mechanical Properties I: Strain, Stress, Stiffness

Ut tensio, sic vis.
(As the extension, so the force).

– Robert Hooke

5.1 Introduction

To begin, it is important to understand that whatever the material, there are two types of deformation that can occur. The first is elastic deformation and the other is plastic[1] deformation. Elastic behaviour is recoverable, plastic behaviour is not. At this point it is useful to note two classes of material which do not conform to the normal rules that will be explored in this chapter. The first are auxetic materials, which will be discussed more fully when we explore the concept of Poisson's contractions, and the second is shape-memory materials, which are able to recover from plastic deformation, typically on the application of heat, without the need to melt or otherwise turn the material back to its constituents and begin again.

Fundamental to an understanding of deformation, and ultimately the failure of materials and structures, is the trifecta of strain, ε, stress, σ, and Young's modulus, E. In the case of a material which deforms elastically, the three are linked by equation (5.1):

$$\sigma = E\varepsilon \tag{5.1}$$

Consider a rubber band. These are notable for their ability to stretch and then return to their original shape and size – elastic behaviour. However, it is possible to over-stretch them, without rupturing them completely. They become noticeably flaccid and are unable to return to their original shape and size – they have been plastically deformed. How much force was required to cause

---

[1] On that basis, it is preferable to avoid talking about plastic materials when one means polymeric.

DOI: 10.1201/9780367822347-5

47

it to stretch? How much did it stretch by when we applied a particular load? In the following sections, we will explore the concepts of strain, stress, and stiffness.

5.2 Strain

In a non-engineering setting, strain and stress are often used interchangeably to denote a mental state.[2] In an engineering context, the two are linked, but have distinct meanings.

Strain is a measure of how much a material or structure (say the rubber band from Section 5.1) will stretch when placed under load. More specifically, strain is the ratio of how much the band stretched, divided by its original length. Typically, the change in length is labelled Δl.[3] By dividing the change in length by the original length ($\Delta l / l$) we get the strain, which is usually given the symbol epsilon (ε):

$$\varepsilon = \Delta l / l \tag{5.2}$$

So, if the rubber band starts with a length of 10 cm and, as we apply load, it stretches by 20 cm then the strain is 20 cm/10 cm, or 2 ε, or 200%ε. From this, it can be seen that strain is dimensionless. Now compare that to a 10 cm long bit of string, it might stretch by just 1 mm for the same applied load, so its strain is just 0.1 cm/10 cm, or 0.01 ε or 1%ε. Table 5.1 provides some examples of equivalence between strain, microstrain, and percent strain.

It is worth noting that there are other measures or definitions of strain. The strain defined above is also referred to as 'engineering strain', or sometimes the Cauchy strain. An alternative to this is to define what is called 'true strain'. Where the engineering strain considers only the total extension at a given point with reference to the original length of the specimen tested, in the case of true strain, the measurement takes into account the change in length occurring as the material is loaded, or, to put it another way, the sum of incremental strains ($\delta \varepsilon_t$), which represent the strain at a given moment, where that incremental strain is defined as the incremental change in length (δl) divided by the length prior to the application of that increment:

---

[2] There is perhaps a nuance that stress sees someone under emotional pressure, whilst strain suggests that emotions are being pulled taught, that a person is being stretched by their commitments.

[3] In physics, the Greek letter delta (Δ) is generally used to denote the change in something. C.f. Δv, the change in velocity due to an accelerating force.

TABLE 5.1

Equivalency in comparative measurements of strain

ε	$\mu\varepsilon$	$\%\varepsilon$
1	1×10^6	100
0.1	1×10^5	10
0.01	1×10^4	1
0.001	1×10^3	0.1
0.0001	1×10^2	0.001
0.00001	1×10^1	0.0001

$$\varepsilon_t = \int \frac{\delta l}{L} \tag{5.3}$$

5.3 Stress

At its simplest, stress is a measure of the overall force, F (or sometimes, P, for pressure) applied to a material, normalised by the area, A, over which it is applied:

$$\sigma = F/A \tag{5.4}$$

In the SI system, force is measured in Newtons (a secondary unit equivalent to 1 kg·m·s⁻², and area is metres squared so that:

$$\sigma = N/m^2 = 1 \text{ Pa} \tag{5.5}$$

Stress may be given either as N/m^2 or as Pa, pascals.[4]

So far, so straight forward: it is in this way that we can compare a steel wire with a thread of spider silk, and determine that the spider silk is stronger, even though it supports a smaller mass before failing. Whilst the mass is smaller, the cross-sectional area of the silk is much smaller than that of the steel wire as the force acting on the spider silk is commensurate with that mass under the influence of gravity. This is a relatively simple process of normalisation. Hence, in designing a structure, we can determine the total force that must be carried and select a material, or a size of a given material, as appropriate.

As an aside, mention strength of materials to most engineers, and they will think of classic textbooks such as those by Timoshenko, which have

---

[4] It is worth noting that 1 MPa is equal to 1 N.mm⁻².

very little to do with materials as described in Chapter 3, but a great deal to do with the behaviour of structures and the various loads that can be applied to them. A good textbook, e.g., Mechanics of Materials (Timoshenko and Gere, 1972), now in its ninth edition (Goodno and Gere, 2020), and no longer with Timoshenko, who died in 1972, listed as an author, is essential when considering some of the more complex issues of applying loads of various types to different structures. It is less useful for understanding how materials fail.

The key thing to remember when considering stress, is that in some senses, it is just a number. Stress only becomes meaningful when we consider that stress in relation to a material, and the effect that it is having on that material. There are two key stresses that are usually defined when considering a material, the yield stress, σ_y, and the ultimate tensile stress, σ_{UTS}. The former is the point at which elastic behaviour ceases and yielding, or plastic deformation, occurs. The latter is the greatest stress that the material can endure before the material enters the endgame of failure. The exact nature of both these events will depend a lot on the nature of the material of interest. Ceramics, for example, yield very little, if at all, and failure occurs almost immediately upon any significant degradation due to the application of load occurring. Some polymers are similarly brittle, whilst others can yield considerably before ultimate failure. Foams yield at comparatively low stresses but reach high compressive strains before final failure mechanisms initiate. Figure 5.1 provides what we might term caricatures of stress–strain behaviours of different materials, in that from these we can recognise the characteristics of some kinds of materials, whilst understanding that we might never see this exact shape in practical testing.

In practice, it is important to understand that the result of a test is not necessarily wrong if it does not conform to the shapes presented in Figure 5.1. Experience suggests that there is a tendency for the novice to assume that they have made a mistake, or that there is something wrong with their sample, or that there is something wrong with the equipment if they do not get a perfectly linear stress–strain response, or a perfectly clear point of yielding, or a perfect post-yield plateau, or a perfectly precise point of failure. All of these may be true:

- It is easy for the sample to be improperly aligned, particularly where a complex specimen is being tested, such as when a composite material is being tested with the fibres at a specific angle to the axis of loading. Good specimen production is critical, but so is attention to detail. Experience and skill can help improve the rate of testing, but good quality data are fundamentally a question of care when setting up the experiment.

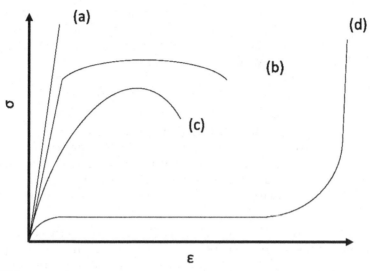

FIGURE 5.1
Examples of stress–strain behaviour: (a) a strong material with a linear relationship between load and extension but which resists deformation, and with a brittle failure mechanism, such as a ceramic; (b) a strong material, again with a linear relationship between load and extension, but which shows some elongation with load, and which yields, with further deformation until failure, such as a metal; (c) a material with a non-linear component to the load–extension relationship, which again shows some yielding behaviour, typical of some polymers; and (d) a material which yields early, but continues to deform in a uniform manner for some time before undergoing further yielding and failure, typical of a foam in compression.

- There will be some level of noise in the output. This will be exacerbated when the test is in the very early stages, and it is important to ensure that a load cell of the correct capacity is selected for the test. Collection of strain data: material type, test geometry, and environmental conditions all need to be considered when selecting strain gauges.
- There are a number of microstructural features that impact on the overall behaviour. For example, Figure 5.1 does not illustrate the kind of stick–slip behaviour that is seen in some steels, whereby immediately post-yield a saw-tooth effect is seen with a small increase in stress followed by a small decrease in stress. Sometimes referred to as discontinuous yielding or plastic flow, this is a function of internal crystal restructuring occurring at a nominally constant level of loading, usually slightly lower than the yield stress, but with some elongation occurring, until the restructuring has concluded, and the material enters the more familiar plateau region.
- Non-linear elasticity is a frequent source of surprise when it is seen for the first time.

It is always worth double-checking your data before continuing with a program of testing: specimens are time-consuming to make and are therefore expensive, even without taking the economic cost of the material into account. This is especially true when you are working on a novel material or processing route and there is a limited number of test specimens available. Before beginning the next test, there are three steps to take:

1. When removing the specimen, look for witness marks from the test protocol. These come in different forms, depending on the test geometry, the material being tested, and the method of holding it secure; however, if the sample has moved during testing, this will be evident in the witness marks, which will become elongated during the test, if the specimen moves.

2. When securing the specimen, look at the fracture surfaces and see if they look as you might expect for the kind of failure you are expecting (ductile or brittle), whether there are any obvious features that show crack growth, or if there are any gross defects.

3. When storing the specimens, whether or not you are planning to study the sample further at this point, act as if you were and ensure that the fracture surface is protected. There is a tendency to want to put the fracture surface back together – *do not do this*, as you will damage the fracture surface further such that it can provide no further information. If it is likely that you will want to undertake chemical analysis of the material after testing, especially of the fracture surface, then wrap the sample in baking foil. Plastic specimen bags will tend to be contaminated with PDMS (polydimethylsiloxane, a release agent). PDMS can cause problems with chemical analysis, as its prevalence and strong chemical signature can result in even small quantities masking chemical features of interest.

If there is any concern with the specimen, specifically the manner of its testing and failure, then this can be addressed before moving onto the next specimen. At this point, before abandoning a test schedule a mental review of the actions so far can go a long way to understanding the results. Have you in fact tested the right specimen? Was there a specific orientation that needed to be tested? Were there any safety or control features that need to be addressed?

It is also worth consulting other practitioners to see if they have any insight. But a fundamental rule, especially when it comes to research, is that just because it does not look like a textbook response, it doesn't mean it's wrong: most textbooks deal in the abstract, in the general and hence examples will tend to be representative, not specific.

5.4 Stiffness

In the 17th century, Robert Hooke stated *"ut tensio, sic vis"*, which means "as the extension, so the force". What he observed was that elastic materials followed a law by which the extension was proportional to the force applied. Different materials extended by different amounts for a given applied force, but whatever the extension of an elastic material, it will double if you double the applied force, triple if you apply three units of force and so on. Of course, there is a limit, where things go PING, which we will come onto in a moment. Hooke was a renowned polymath with a range of interests, including the microscopic world and his architectural ambitions for a London being rebuilt after the Great Fire of London, not to mention his long-running and ultimately career-ruining feud with Sir Isaac Newton. This perhaps explains why he didn't take this to its logical conclusion, i.e., that there is a relationship between stress and strain, and that when an elastic material is extended in this way, the resistance to extension, as defined by this relationship between stress and strain can be defined as the stiffness. This relationship was first defined by Leonard Euler, in the early 18th century, but, for reasons which do not concern the present text, we talk about the Young's modulus,[5] E, of a material: $E = \sigma/\varepsilon$. The Pyramid of Tests sets out a hierarchy of testing which starts with materials at the bottom and finishes with structures at the top.[6] Humans have been building things in a serious way for at least 5000 years – if you start with megalithic stone structures. In a mere six hundred or so years we go from 'simple' structures, with one stone lintel placed on top of two uprights, to the pyramids of Egypt, and on to the large-scale structures of ancient China and Mesoamerica. These were built with the materials available and were designed on the basis of tried and tested methods: empirical design. Archaeologists are still trying to determine how some effects were achieved with what is known of the tools available at the time. Empirical methods were still employed during Europe's Medieval period when many of the great cathedrals and castles were built, sometimes over the course of hundreds of years.

But it is only within the last hundred or so years that we have moved towards predictive design, based on models, and it is only in the last fifty or

---

[5] As in Thomas Young, another polymath, who in addition to many other feats helped to translate Egyptian Hieroglyphs using the Rosetta Stone. His paper on the subject of E dates to the very early 19th century.

[6] Jerry Lord of Boeing has suggested that there should be a second pyramid that should be considered at the same time. Whilst not entirely relevant to the current context, it is worth noting for completeness. This second pyramid considers the validation of the manufacturing process.

so years that this design process has been dependent on the materials available, especially those of a synthetic nature.

5.5 Poisson's Contractions, and an Aside on Auxetic Materials

As discussed in Section 5.2, when we place a material under load, the material extends. If the material is below the yield stress, then the extension will be elastic, and recoverable. If we go beyond the yield stress, then plastic deformation will occur. But that is not the whole story. The volume of material will stay constant, until plastic deformation occurs. The logical inference, then, is that the volume must change form to accommodate the elongation: as the direction parallel to the loading stretches, the direction perpendicular contracts (Figure 5.2a). This is called a Poisson's contraction, and the extent of this can be calculated using the Poisson's ratio, v:

$$v = -\frac{\varepsilon_{transverse}}{\varepsilon_{axial}} \tag{5.6}$$

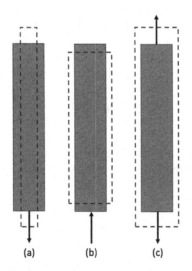

(a)   (b)   (c)

FIGURE 5.2

Poisson's 'reactions': (a) in tension, we would normally expect to see the sample elongate in the direction of load, whilst the transverse direction contracts; (b) in compression we observe the opposite effect in that the specimen reduces in length in the direction of the applied load and the transverse width increases; and (c) in an auxetic material in tension, all dimensions increase.

The reverse is also true: if we place the material under compression, the material will expand in the perpendicular direction(s).[7] When it comes to a physical test, in the case of a compressive test, it is important to consider the level of friction between the sample and the platens constraining the specimen. If the friction is too great, the material will bulge, in a form called barrelling (Figure 5.3). If the friction is too small, then there is a danger that the sample will slip out from between the platens.

Typically, the Poisson's ratio of a material lies somewhere between 0.3 and 0.5, but some materials, usually engineering foams and the like, can have a ratio approaching 0, which is to say that there is very little in the way of Poisson's reaction.[8] In terms of extreme behaviours, this brings us to the case of auxetic materials. These unusual materials actually increase in all directions when placed in tension (Figure 5.3c), or contract in all directions when compressed in one, because of the internal structure. Here again is an example of a mesostructure, where internal structuring of the material at a level between the micro- and macrostructures is such that instead of the normal process of deformation, the material opens (or closes) with the tensile (or compressive) load.[9] For a more in-depth consideration of auxetic materials, see Lim (2015).

5.6 Summary

The relationship between stress, strain, and Young's modulus is a fundamental one. It is important to remember that at different times, a given design may be stress limited or it may be strain limited. That is to say, the design may be required to carry a particular load, which will require the design to have a certain cross-section. If the cross-section is greater than can be accommodated

---

[7] To avoid confusion between orthogonal directions, here the terms axial, i.e., parallel, and transverse, i.e., perpendicular, are used, with reference to the loading direction. (Some conventions would align the x direction with the direction of loading, others the z direction.) The default convention is also to consider Poisson's contractions as a two-dimensional problem, with the through thickness direction effectively ignored. This is not generally a problem. Where a flat specimen is considered, the relative through thickness contraction will be comparatively small, and where the material is isotropic, at least in the directions perpendicular to the applied load, then the Poisson's ratio will be the same in both transverse directions. Things become more complex when there is a question of anisotropy to account for, i.e., there will be a different Poisson's ratio for each transverse direction.

[8] This is particularly the case with foams placed in compression, where the cellular structure tends to collapse in on itself rather than distorting in other ways. See also Section 4.6.

[9] Whilst there are some materials that display auxetic behaviour, the majority of examples are not really materials, in a sense, but are structures, albeit at a small scale. In section, they are much like a foam, with volumes of open space. The space is bounded by material, which is not necessarily special of itself, but rather by the way it is shaped.

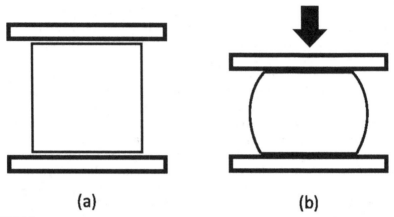

(a)   **(b)**

FIGURE 5.3
The issue of barrelling. When a sample is placed in compression (a), instead of compressing uniformly throughout the volume, the sides bulge (b). This arises due to friction between the platen and the sample, which prevents the sample from accommodating the change in shape arising from the applied load.

due to other constraints (such as other components, or the footprint or volume available for the component) then it may be necessary to select a stronger material that can carry the specified load within a smaller component. Alternatively, the material initially selected may be strong enough to fit within the volume available, but under load it may deflect too much. It has not failed in the sense that the material has not plastically deformed; still the material has failed in the sense that it is unable to do the job that is required of it. Hence, we do not need to select a stronger material, but rather one that is stiffer, and better able to resist the load applied.[10]

References

Goodno, B.J. and Gere, J.M., 2020. *Mechanics of materials*, 9th Edition. Cengage Learning.

Lim, T.C., 2015. *Auxetic materials and structures* (pp. 55–56). Springer.

Timoshenko, S. and Gere, J.M., 1972. *Mechanics of materials*, 1st Edition, D. Van Nostrand Company.

[10] This presents the bias of a materials scientist. A designer might default to redesigning the component or structure, especially if there is a requirement to use a particular material. It will of course depend on the overall constraints of the project as a whole, including budget, the time available to develop the design, environmental concerns, as well as the footprint/volume available.

Recommended Reading

In addition to the textbook that is required reading for Engineering students around the world, Timoshenko wrote up lecture notes for his course on the history behind the development of fundamental engineering theory:

Timoshenko, S., 1983. *History of strength of materials: With a brief account of the history of theory of elasticity and theory of structures.* Courier Corporation.

6

Mechanical Properties II: Toughness and Fracture Toughness

The mirror crack'd from side to side,
The curse is come upon me cried,
The Lady of Shalott.

– Alfred, Lord Tennyson

6.1 Introduction

It can be tempting to think that so long as we know the strength of a material, its strain to failure, and its Young's modulus, then we know everything we need in order to design structures using that material. In some senses that is true, but it only deals with one facet of the design process. Two further mechanical properties that need to be considered are the toughness and the fracture toughness. Together, these properties form the core of fracture mechanics.

Fracture mechanics is the study of cracks, specifically how big does a crack need to be in order to be a problem. There are two aspects to consider here, the amount of energy required to create two new surfaces (the toughness), and the resistance to crack growth (the fracture toughness). The latter in particular is a function of the material, the sharpness of the crack and its orientation with respect to applied loads, the load applied, the geometry of the component and the position of the crack in it, and various environmental conditions. The former is more related to the nature of the material itself, and its ability to deform.

Fracture mechanics is divided into various sub-groupings, the two most important being linear elastic fracture mechanics (LEFM) and elastic–plastic fracture mechanics (EPFM).

DOI: 10.1201/9780367822347-6

6.2 Background to Fracture

In order to understand fracture mechanisms, it is necessary to take a step back and consider the event of fracture, in other words, the effect that is observed (fracture) rather than the cause (fracture mechanisms). In this respect, a good definition of fracture is provided by Andrews (1968):

> *By fracture we mean the creation of new surfaces within a body...The definition as it stands does not specify how the creation of new surfaces is effected...*

Andrews (1968) also specifies that this definition means that whilst some words (e.g. 'rupture') could be used interchangeably with fracture, 'failure' is a word that simply indicates that a component or test piece can no longer be used for its original purpose, with additional information being required to specify how failure occurred. Equally, 'cracking' is not interchangeable with fracture since whilst it clearly involves fracture, there are other forms of fracture which have nothing to do with cracking. In fact, Andrews (1968) proposes six separate forms of fracture:

I.   Fracture under direct loading
II. Crack propagation
III. Fatigue
IV. Creep fracture (static fatigue)
V.   Wear (abrasion)
VI. Environmental (stress corrosion cracking)

Andrews (1968) treated crack propagation as a separate issue so that he could raise the particular points of notch sensitivity and high-speed fracture. Notch sensitivity is an extremely important factor in some materials (which are in turn termed notch sensitive) since the concentration of stress caused by a sharp notch or crack can cause failure at relatively low loadings.[1] Of the above list, the most relevant forms of fracture to this study are I and II. Crack propagation contributes to fracture under direct loading, and so this study is in fact interested in a combination of I and II rather than a separate analysis of each. However, this list is perhaps not helpful, in that it returns

---

[1] Recall the point made in the previous chapter regarding the difference between stress (a general term for the effect of an applied load on a given surface area) and strength (a specific stress for a given material at which some significant behaviour occurs, such as plastic yielding, the yield strength, or complete failure, the ultimate tensile strength). For some materials, the presence of some kind of degradation will just lead to a reduction in the overall area that can carry the load. For others, the presence of a crack will lead to a stress state at the crack tip which will exceed the strength of the material at much lower loads than might be expected.

us to the question of cause and effect. We can observe a failed specimen and attribute failure to a particular mechanism, but the failure observed may not necessarily be a fracture: the generation of a crack and its evolution to a critical crack can be attributed to a specific mechanism, but the mechanism in question can also cause other forms of failure. For example, abrasion may lead to the formation of a crack, or it may lead to a more stable thinning of the component in a specific location which means that the component fails, in that it is no longer performing its proper function, or which may lead more extreme issues in due course.[2]

6.3 Pre-Griffith Concept

Lawn (1993) summarises the state of knowledge in the early part of the 20th century (i.e., prior to Griffith's work) as being based on the concept of any given material having a particular strength. Hence, Griffith (1920) states:

> *"According to these hypotheses rupture may be expected if (a) the maximum tensile stress, (b) the maximum extension, exceeds a certain critical value. Moreover, as the behaviour of the materials under consideration, within the safe range of alternating stress, shows very little departure from Hooke's Law, it was thought that the necessary stress and strain calculations could be performed by means of the mathematical theory of elasticity...*
>
> *"...Thus, on the maximum tension hypothesis, the weakening of, say, a shaft 1 inch in diameter, due to a scratch one ten-thousandth of an inch deep, should be almost exactly the same as that due to a groove of the same shape one hundredth of an inch deep...*
>
> *"...These conclusions are, of course in direct conflict with the results of alternating stress tests...*
>
> *"...To explain these discrepancies, but one alternative seemed open. Either the ordinary hypothesis of rupture could be at fault to the extent of 200 or 300 per cent., or the methods used to compute the stresses in the scratches were defective in a like degree."*

---

[2] For example, leakage at the joint of a buried water main can lead to a hard object, such as a pebble, in a soft object, such as soil or backfill, rotating, causing the pipe to wear and a hole to form. Technically the pipe has failed twice over at this point, as water is escaping from the joint, as well as through the hole created by the wear caused by the rotating pebble. However, the pipe is still able to perform its primary function with a greater or lesser loss of efficiency. However, the greater volume of water now escaping can lead to the pipe becoming undermined as the pipe's support is washed away. In due cause, a more significant failure will occur if the problem remains unidentified for any length of time.

Hence, the common theories, prior to 1920, regarding the rupture of materials had started to come into disrepute, due to non-reproducibility of results and variations in results developing from variations in test conditions.

The 'Molecular Theory of Strength' (based on the energy required to separate two atoms so that they no longer interact) suggests that the stress at fracture (or fracture strength), σ_{max}, is given by:

$$\sigma_{max} = \sqrt{\frac{E\gamma}{b_0}} \tag{6.1}$$

where E is the Young's modulus, γ is the surface tension, and b_0 is the interatomic spacing. This typically leads to predicted strengths of ~$E/10$. In practice it was found that the stress at fracture was much lower, at ~$E/1000$. This is approximately two orders of magnitude lower than predicted. Griffith (1920) was the first to propose a theory to explain this discrepancy.

Griffith built on early work by Inglis (1913) who provided an analysis which showed the effect of stress concentration by elliptical cut outs[3] in uniformly stressed plates. Whilst the mathematical treatment provided by Inglis (1913) was significant, it begged the question of why large cracks were more detrimental than smaller cracks despite the apparent size-independence of the stress concentration factor.

6.4 The Griffith Concept

Griffith's (1920) contribution to the field of fracture mechanics was two-fold. Firstly, whilst Inglis (1913) proposed that a crack could be treated as an almost infinitely thin ellipse (which it can), with the associated stress concentration at the crack tip, it was Griffith who proposed that materials might contain flaws or defects for which the Inglis solution might be applicable. By noting the discrepancy observed between real and theoretical fracture stresses in brittle materials, Griffith determined that in specimens that had no readily observable cracks or surface damage, a defect of microscopic size and the associated stress concentration would account for the discrepancy observed (Knott, 1973; Lawn, 1993).

Secondly, Griffith (1920) proposed a thermodynamic solution to the problem of modelling crack growth. Crack growth is dependent upon the strain energy, U. U is the sum of the strain energy of a crack tip field, U_E (as a

---

[3] By making one axis much smaller than the other, the ellipse may be assumed to be a mathematical model of a crack.

result of an applied stress, σ, equation [6.2]) and the surface energy of a crack of length 2a, US (where γ is the surface energy per unit area, equation [6.3]).

$$U^E = \frac{\sigma^2 \pi a^2}{E} \tag{6.2}$$

$$U^S = 4\gamma a \tag{6.3}$$

Under a given stress the total energy reaches a maximum at a particular value of crack length. Therefore, crack growth will occur when the strain energy release rate is equal to or greater than the rate of change of surface energy with crack length:

$$\frac{dU^E}{da} \geq \frac{dU^S}{da} \tag{6.4}$$

By considering the change in the energy of the system with crack length it is possible to show that for crack growth to occur:

$$\frac{dU}{da} = \frac{dU^S}{da} - \frac{dU^E}{da} = 0 \tag{6.5}$$

Substituting equations (6.2) and (6.3) into equation (6.5), and differentiating, it may be shown that:

$$\sigma\sqrt{\pi a} = \sqrt{2\gamma E} \tag{6.6}$$

and since the critical strain energy release rate, G_c, is equal to twice the surface energy, γ:

$$G_c = 2\gamma \tag{6.7}$$

and introducing the critical stress intensity factor, K_c, which is equal to $\sigma\sqrt{\pi a}$, equation (6.6) may be written in the form:

$$K_c = \sqrt{G_c E} \tag{6.8}$$

Griffith's concept was found to be generally applicable to 'elastic' systems with one caveat: being based on an energy balance utilising the Second Law of Thermodynamics, the theory as it stands predicts that if the energy balance is tipped one way then a crack will extend, whilst if it is tipped the other way, for example when the load is removed and the sample becomes unstressed, then the crack formed by the previously applied stress should 'heal' or close up. The healing of cracks has been reported, and there are very good reasons why it can potentially occur. However, it is important to maintain a healthy scepticism when considering a report of self-healing: there must be a compelling reason for the cracking process to be undone and for bonds to reform. Simply juggling energy flows is not sufficient. In this context, it is also important to acknowledge that the term self-healing is also used in materials science to denote a material that has been designed to incorporate a repair mechanism.[4]

Also, the Griffith construct is based on energy changes between an initial and final state and so represents only a necessary condition for fracture. As it does not deal with the actual mechanism of the fracture concerned there may be other criteria which must also be fulfilled for fracture to occur.

6.5 Quasi-brittle Fracture

To make the Griffith concept more generally applicable, one of the first steps to take into account is plastic yielding at the crack tip. This was done, independently, by both Orowan (1955) and Irwin (1958). They determined that if a specimen were in a state of plane strain,[5] only limited plastic deformation could take place, and therefore the plasticity would be allowable for linear elastic fracture mechanics (LEFM) to be applicable. Under these circumstances it is possible to reassess G_c; equation (6.7) can be rewritten to consider the surface energy, γ_s, and the energy absorbed by the sample by plastic deformation, γ_p:

$$G_c = 2\left(\gamma_s + \gamma_p\right) \qquad (6.9)$$

Irwin (1958) presented a criterion for the minimum specimen thickness for the specimen to be in plane strain. This has subsequently been revised, and here it is presented in the more general form, as it is found, for example, in ASTM D5045-14R22 (2022):

---

[4] Biological organisms such as the human body are excellent examples of self-healing mechanisms, and there are certain cements (and derived concretes) and composite materials which are designed to repair cracks that occur during operation. However, such repair mechanisms are not self-healing within the context of the damage of a crack being undone simply by unloading the component.

[5] The concept of plane strain, and indeed plane stress, is examined more fully in Chapter 8.

$$B > 2.5\left(\frac{K}{\sigma_y}\right)^2 \qquad (6.10)$$

6.6 Linear Elastic Fracture Mechanics

The fundamental equation for LEFM is:

$$K = \sigma\sqrt{\pi a} \qquad (6.11)$$

where K is the stress intensity factor generated for a given applied stress, σ, and a is half the length of the crack under consideration. When K is K_c, the critical stress intensity factor, then this is a property of a given material.[6] However, equation (6.11) is something of special case – it considers a crack in a semi-infinite plate, which is to say that there are various factors of constraint that we don't have to think about. A generic form of equation (6.11), which begins to take into account various geometric factors is:

$$K = Y\sigma\sqrt{a} \qquad (6.12)$$

The factor of π is subsumed into the geometric term, Y, where Y can be modified to take into account, for example, the ratio of the length of the crack with respect to the width of the plate (a/w) (see e.g., Hertzberg et al., 2020), the angle of the crack with respect to the applied load, and the overall shape of the component (see e.g., Boresi and Schmidt, 2002). A seminal example of the development of a geometrical factor for a specific application is Newman and Raju's (1980), calculated for a crack with a specific profile and location in a thin-walled pressure vessel.

---

[6] As with stress and strength, K represents a state that exists when a load is applied to a sample that contains a crack. K_c is the material property, which means that if we know K_c and either the applied stress or the crack length, then we can determine the other. If we know that there will be a particular load applied to a structure, then we can determine the length of the critical crack. Conversely, if we have identified a crack, perhaps through NDE, or even simple observation, and we can determine the length of the crack, then if we also know the K_c for the material, then we can calculate the maximum stress that can be applied before failure, and act accordingly.

6.7 Elastic–Plastic Fracture Mechanics: The J-integral

Elastic–plastic fracture mechanics deals with the case of materials that do not satisfy the requirements of LEFM, but instead show more significant plastic deformation around the crack tip. Plastic behaviour is often associated with yielding (the stress at which a material will no longer behave elastically) but polymers are especially likely to show some degree of material non-linearity at comparatively low stresses. This is particularly true when considering the failure of a material due to crack growth. Plastic failure is non-reversible but can often absorb large energies during crack growth.

The J-integral was proposed as a parameter to deal with small-scale yielding in elastic–plastic materials (Rice, 1968). The J-integral is path-independent:

$$J = \int_{\Gamma}^{J} \left(W_e \, dy - T . \frac{\delta u}{\delta x} ds \right) \qquad (6.13)$$

where x and y are rectangular coordinates normal to the crack front,[7] ds is an element of arc length along the contour Γ, T is the stress vector acting on the contour, u is the displacement vector, and W_e is the strain energy density. This is illustrated in Figure 6.1.

Rice (1968) showed that J could be determined from the difference in the load displacement behaviour of two specimens of the same material which differ only in initial crack lengths a and $a+\Delta a$.

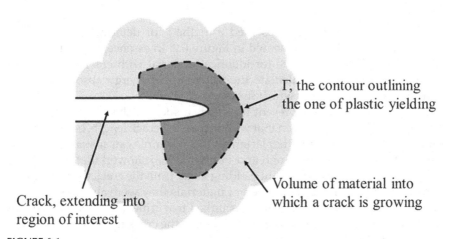

Γ, the contour outlining the one of plastic yielding

Volume of material into which a crack is growing

Crack, extending into region of interest

FIGURE 6.1
Simplified illustration of the application of the J-integral.

Source: Adapted from Rice (1968).

---

[7] y being perpendicular to the crack surface.

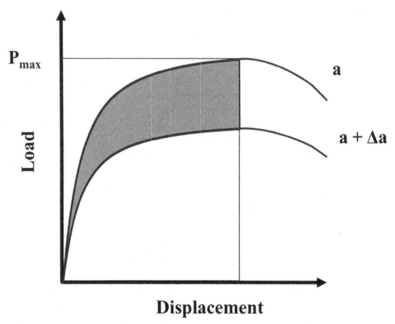

Displacement

FIGURE 6.2
Determination of -ΔU. After Begley and Landes (1972).

$$J = \frac{-\Delta U}{\Delta a} \qquad (6.14)$$

Begley and Landes (1972) presented a method for determining $-\Delta U$ from experimental results, as presented in Figure 6.2. In essence, the difference in the load–displacement curve for identical material with cracks of different lengths can be used to calculate the difference in energy absorbed by the material until the point at which the crack grows.[8]

In fact, many samples present a sequence such as that demonstrated in Figure 6.3. Here it can be seen that with increasing crack length, both the maximum load and maximum displacement that a sample can sustain decreases. Agarwal et al. (1984) and Singh and Parihar (1986) showed that this type of behaviour was due to significant yielding away from the crack tip for samples with short initial cracks. For different materials they were able to show that yielding away from the crack tip was linked to a critical crack length so that for deeply notched specimens only yielding at the crack tip was observed (a/W = 0.35 for Agarwal et al., 1984, and a/W = 0.34 for Singh and Parihar, 1986). Kim and Kim (1989) reported that they could not find such a critical crack

---

[8] $-\Delta U$ may also be calculated from a pair of samples with the same maximum load but where the displacement due to $a+Da$ is greater than that due to a.

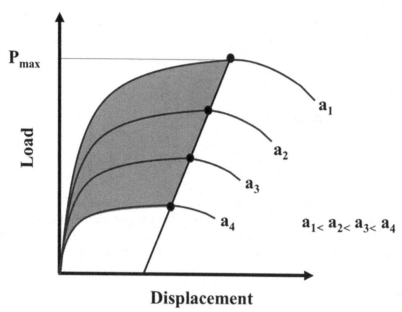

FIGURE 6.3
Evaluation of $-\Delta U$ for the case where crack propagation occurs at decreasing load and decreasing displacement for samples with increasing crack length. After Kim and Joe (1988).

length but stated that at a value of a/W = 0.4 the relationship between energy dissipation away from the crack tip and the initial crack length became linear.

Kim and Joe (1987, 1988) presented a 'locus' method which seeks to separate the contributions to the total deformation energy required for the propagation of the initial crack for a sequence of samples with increasing initial crack length. The locus method seeks to provide a 'partition' between the energy absorbed at the crack tip and that absorbed away from the crack tip.

Kim and Kim (1989) have used this method to assess a 'short glass fibre reinforced thermoplastic polyester' and found that JC and GC were in good agreement since the material showed only a minor plastic component to its behaviour. Their work shows some unexpected results, however. For example, in Figure 6.3, it is clear that the displacement due to a sample decreases with increasing crack length. Kim and Kim (1989) present samples where a similar relationship of four crack lengths has a4 with a displacement greater than that of a2 and a3.

Rice et al. (1973) showed that J could be determined using a single specimen. In this method, the displacement is partitioned into two components, the elastic, δ_{el}, and the plastic, δ_{pl}, and hence J is made up of two components, the elastic, J_{el}, and the plastic, J_{pl}, contributions to fracture:

$$J = J_{el} + J_{pl} \tag{6.15}$$

J_{el} is essentially the LEFM critical strain energy release rate or toughness, G, and may be determined by:

$$J_{el} = G = \frac{P^2}{2B}\frac{dC}{da} \qquad (6.16)$$

whilst J_{pl} may be determined by:

$$J_{pl} = \frac{1}{B(W-a)}\left(2\int_0^{\delta_{pl}} Pd\delta_{pl} - P\delta_{pl}\right) \qquad (6.17)$$

In the preceding equations, P is the maximum experimental load, B is the specimen thickness, dC/da is the differential of the function that describes the relationship between C, the compliance of the sample, and a, the crack length. Atkins and Mai (1988) showed that equation (6.14) could be simplified to:

$$J_{pl} = \frac{2A^*}{B(W-a)} \qquad (6.18)$$

where A^* is the area described by the relation between stress and strain and a straight line between 0 and the maximum load, as illustrated in Figure 6.4. As may be imagined, this analysis is dependent upon specimen geometry. In equation (6.18), a single deep notch is considered. Equation (6.18) is suitable

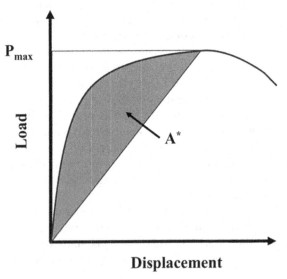

FIGURE 6.4
Demonstrating the derivation of A*.

for double-edge or central-notched specimen geometries, for example; it is not suitable for compact tension or three-point bend specimens.

6.8 Summary

This chapter has considered material properties of toughness and fracture toughness, and the impact of defects on mechanical performance, where those defects act to cause stress concentration. When undertaking testing of materials to determine toughness and fracture toughness, it is important to ensure that the correct standard is selected for the materials being tested, and that factors such as whether the material is in plane strain or plane stress have been taken into account. It is also important to consider the structure of the material to be tested and whether there are other factors that may impact on the specimen to be tested. For example, if testing material that contains particulate reinforcement, the minimum depth of the specimen will need to be several times the diameter of the reinforcement to ensure that the reinforcement does not sit proud of the surface and does not have an undue effect on the matrix, for example, constraining its ability to deform. Hence, whilst there are fundamental principles to be observed in every test, there are caveats and modifiers for every material to be tested.

References

Agarwal, B.D., Patro, B.S. and Kumar, P., 1984. "J integral as fracture criterion for short fibre composites: An experimental approach". *Engineering Fracture Mechanics*, 19, pp.675–684.

Andrews, E.H., 1968. *Fracture in polymers*. Oliver and Boyd.

ASTM D 5045 - 96, 1996. "Standard test methods for plane-strain fracture toughness and strain energy release rate of plastic materials", *Annual Book of ASTM Standards* (pp.325–333). ASTM, Philadelphia.

Begley, J.A. and Landes J.D., 1972. "The J Integral as a fracture criterion", *Fracture toughness, Proceedings of the 1971 National symposium on fracture mechanics, Part II, ASTM STP 514* (pp.1–20). American Society for Testing and Materials.

Boresi, A.P. and Schmidt, R.J., 2002. *Advanced mechanics of materials*, Sixth Edition. John Wiley & Sons.

Griffith, A.A., 1920. "The phenomena of rupture and flow in solids". Philosophical Transactions of the Royal Society, A221, pp.163–198.

Hertzberg, R.W., Vinci, R.P. and Hertzberg, J.L., 2020. *Deformation and fracture mechanics of engineering materials*, Sixth Edition. John Wiley & Sons.

Inglis, C.E., 1913. "Stresses in a plate due to the presence of cracks and sharp corners". *Transactions of the Institute of Naval Architects*, 55, pp.219–241.

Irwin, G.R., 1958. "Fracture", in *Encyclopaedia of Physics* (Vol. VI, pp.551–590), ed. S. Flügge, Springer-Verlag OHG.

Kim, B.H. and Joe, C.R., 1987. "A method to determine the critical J-Integral value independent of initial crack sizes and specimen lengths". *International Journal of Fracture*, 34, pp.R57-R60.

Kim, B.H. and Joe, C.R., 1988. "Comparison of the locus and the extrapolation methods that determine the critical J-Integral in the presence of remote energy dissipation". *Engineering Fracture Mechanics*, 30, pp.493–503.

Kim, B.H. and Kim, H.S., 1989. "Fracture characterization of short glass fibre reinforced thermoplastic polyester by the J-Integral". *Journal of Materials Science*, 24, pp.921–925.

Knott, J.F., 1973. "Fundamentals of Fracture Mechanics", Butterworths.

Lawn, B.R., 1993. *Fracture of Brittle Solids*, Second Edition. Cambridge University Press.

Newman Jr, J.C. and Raju, I.S., 1980. Stress-intensity factors for internal surface cracks in cylindrical pressure vessels. *Journal of Pressure Vessel Technology*, 102, pp.343.

Orowan, E., 1955. "Energy criteria of fracture". *Welding Research Supplement*, 34, pp.157–160.

Rice, J.R., 1968. "A path independent integral and the approximate analysis of strain concentration by notches and cracks". *Journal of Applied Mechanics*, 35, pp.379–386.

Rice, J.R., Paris, P.C. and Merkle, J.G., 1973. "Some further of J-Integral analysis and estimates" in *Progress in Flaw Growth and Fracture Toughness Testing (ASTM STP 536)* (pp. 231–245). American Society for the Testing of Materials.

Singh, R.K. and Parihar, K.S., 1986. "The J-integral as a Fracture Criterion for Polycarbonate Thermoplastic". *Journal of Materials Science*, 21, pp.3921–3926.

7

Fatigue, Wear, Creep, and Note on Corrosion

It is better to wear out than rust out.

– Proverb

7.1 Introduction

Where strength, Young's modulus, toughness, and fracture toughness are material properties, fatigue, wear, and creep are a material's response to sub-critical loads. Corrosion is something of a special case, a form of degradation that reduces the carrying capacity of a structure. The thing all four have in common is that failure does not occur because the structure has been bought to the ultimate tensile strength of the material in an instant, but rather because, through continual use and exposure to the environment, the amount of material that is capable of carrying load is reduced.

7.2 Fatigue

If we have had a long day, we might say that we are tired, or perhaps fatigued. There are tasks that we perform easily, but at the end of such a day, or if we repeat them a number of times, we might find ourselves making mistakes, or perhaps suffering physically more than we do if we complete the tasks when fresh. The greater the number of repetitions, the greater the impact.

Materials and structures, in some senses, are no different. Recalling our determination of the strength of a material, this property relates to the value at which an applied load causes certain changes to occur in the microstructure of the material (yield strength) or which causes the destruction of the sample tested (ultimate strength). But what happens if we load the material to some point short of an upper limit? What happens if we repeat a test a number of times? Like us repeating an exercise a number of times, the material becomes fatigued, and damage that was non-critical when the sample was tested short

DOI: 10.1201/9780367822347-7

of complete failure for one test becomes a problem when we repeat the test. How much of a problem will depend on the number of times that the specimen goes through a loading cycle.

Some things, therefore, to keep in mind: every time a vehicle goes over a bridge, every time an aeroplane takes off and lands, every time a car goes over a speed bump, part of the structure undergoes a loading cycle. Have a think about things that you do repeatedly in your daily life: putting on the kettle, switching on a light, opening a door…

The key issue with fatigue is determining the number of loading cycles that a material can survive at a given applied stress. Clearly, a material cannot undergo more than one cycle at a critical stress before yielding or catastrophic failure prevent the structure from being used again. It is necessary to develop an understanding of the behaviour of the material for different applied loads, and determine a model that can be used to predict behaviour, for a given load cycle (Figure 7.1).

Whilst a single curve can be defined, a distinction is usually drawn between low-cycle and high-cycle fatigue, with different failure mechanisms being observed, depending on the balance of applied load and the number of cycles to failure. For example, Murakami and Miller (2005) describe the impact of high-stress, low-cycle fatigue on the fracture ductility of materials. Samples with pre-drilled holes became more susceptible to crack propagation after low-cycle fatigue, no matter how small the defects already present.

Commonly, the Coffin-Manson equation is used to describe low-cycle fatigue. As it is easy to misinterpret this equation, it is often helpful to consider it as one equation formed of two parts:

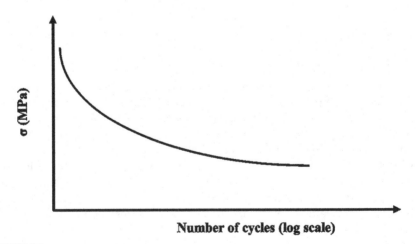

Number of cycles (log scale)

FIGURE 7.1

Indicative stress–number (S–N) curve demonstrating the relationship between applied stress and the number of cycles that a structure can go through at a given load before failure.

$$\frac{\Delta \varepsilon_T}{2} = \frac{\Delta \varepsilon_e}{2} + \frac{\Delta \varepsilon_p}{2} \tag{7.1}$$

i.e., that the total strain amplitude, $\dfrac{\Delta \varepsilon_T}{2}$, is comprised of an elastic strain amplitude, $\dfrac{\Delta \varepsilon_e}{2}$, and a plastic strain amplitude, $\dfrac{\Delta \varepsilon_p}{2}$. The elastic part of the equation is defined as:

$$\frac{\Delta \varepsilon_e E}{2} = \sigma_a = \sigma'_f \left(2N_f \right)^b \tag{7.2}$$

where σ_a is the stress amplitude, σ'_f is the fatigue strength coefficient, an empirical constant defined by the stress intercept at 2N = 1, $2N_f$ is the number of strain reversals to failure,[1] and b is the fatigue strength exponent. This latter is a material property, but varies considerably. Similarly, the plastic part is:

$$\frac{\Delta \varepsilon_p}{2} = \varepsilon'_f \left(2N_f \right)^c \tag{7.3}$$

where $\dfrac{\Delta \varepsilon_p}{2}$ is the plastic strain amplitude, ε'_f is the fatigue ductility coefficient, an empirical constant defined by the strain intercept at 2N = 1, and c is the fatigue ductility exponent. This latter is a material property, but will generally lie between –0.5 and –0.7.[2]

Hence:

$$\frac{\Delta \varepsilon_T}{2} = \frac{\sigma'_f}{E} \left(2N_f \right)^b + \varepsilon'_f \left(2N_f \right)^c \tag{7.4}$$

It is tempting to try to simplify the equation by assuming that $\dfrac{\sigma'_f}{E} = \varepsilon'_f$, but as the two are defined on the basis of separate graphs, this is not an appropriate step to take.

Whilst it is normal to think in terms of the fatigue cycle as being a sinusoidal wave function (Figure 7.2), other wave shapes are used in certain circumstances, and it is important to check what shape is required for a

---

[1] That is, twice the number of cycles, N_f, to failure, because there are two strain reversals per cycle.

[2] There is no correlation between b and c, although b is always smaller, and the exact ratio varies between approximately 1/10 and 1/4.

specific application. It is also important to note that just because a particular wave form is programmed into the control software, what the specimen is subjected to may be different, and non-uniform. The most common reasons for this to occur are:

- A high cyclic rate is rate is required;[3]
- The load applied is close to the maximum capacity of the frame;
- The difference between maximum and minimum applied load is very small; or
- The test frame is worn or in need of service.[4]

As a result of any of the factors above, or a combination, there are three key issues to watch out for:

- Failure to reach the maximum or minimum loads before the next cycle starts;
- 'Smearing' of the peak, where there is some flattening of the wave-peak; and
- Asymmetry of the wave-form where one side of the wave takes longer and the other is slower.

Figure 7.2 represents the typical features of a fatigue curve, but one where the peak tensile stress and the peak compressive stress are equal. In practice, such a cycle, where the specimen undergoes full reversal, is at the more difficult end of testing, and is not the typical fatigue loading cycle to be considered. More normally, fatigue will be carried out in a fully tensile or fully compressive manner, which is more representative of how most structural members will behave in operation. Hence, the cycle is usually defined by a ratio of peak and minimum applied stress, R, which will usually equal 0.1, although other values of R may be used on occasion:

$$R = \frac{\sigma_{min}}{\sigma_{max}} \tag{7.5}$$

Considering, Figure 7.2 again, another feature that we can see that has been idealised is that the fatigue cycle is uniform over the whole number of cycles considered. In terms of testing, this is generally the case; in practice, over the

---

[3] What constitutes high in this case will depend on the capacity of the test frame and its mode of operation. An electrodynamic frame can respond more rapidly than a hydraulic frame, for example.
[4] A particular issue with hydraulic test frames is the potential for the oil in the power-pack to become overheated.

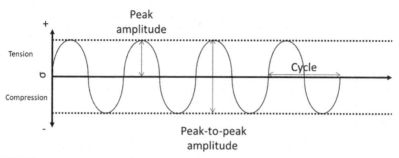

FIGURE 7.2

Key features of fatigue. A fatigue cycle can be defined from any starting point, and this point will be seen at three points in any cycle, at the beginning of the cycle, at the end of the cycle, and at some point in the middle, depending on exactly which starting point is chosen. The exception to this is if the maximum or minimum point of the cycle is chosen as the starting point.

lifetime of a component, it may undergo a range of fatigue cycles that are non-uniform. For example, we might expect the rest state to stay the same, barring any significant changes to the material over time, but will the structure return to the rest state before a new fatigue cycle starts? Possibly, most of the time, but on occasion we may see a flurry of activity which means that the minimum tensile load is greater than normal. Similarly, if the loading is more gentle for a time, the peak load will be lower for a time. Some research has been undertaken to investigate the impact of mixed fatigue behaviour on a structure, and to see if this has greater significance than if the structure were subjected to these independently (see, e.g., Sonsino, 2007).

A final aspect to consider with respect to fatigue is around fatigue crack growth, which is modelled using the Paris (sometimes referred to as the Paris-Erdogan) Law:

$$\frac{da}{dN} = C\left(\Delta K\right)^m \tag{7.6}$$

where da/dN is the change in crack length in a single fatigue cycle, ΔK is the difference between the maximum and minimum stress intensity factors:

$$\Delta K = K_{max} - K_{min} \tag{7.7}$$

C and m are determined experimentally, and are dependent on various factors: when building a model for a specific situation, it will be necessary to undertake multiple series of tests to identify modifiers to C which address various contributing factors such as temperature. Materials may also exhibit different fatigue crack growth behaviours when in different environments, such as whether they are submerged in clean, brackish, or salt water. It

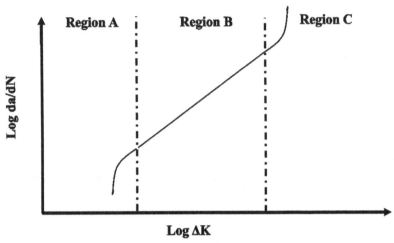

FIGURE 7.3

Illustration of the crack growth rate vs. stress intensity range. At very low (Region A) or very high (Region C) ΔK, crack growth is unstable. The Paris Law only applies to stable crack growth (Region B).

should also be noted that the Paris Law only applies to stable crack growth (see Figure 7.3).

7.3 Wear

The focus of this chapter is very much on the aspects of failure that relate to the mechanical properties of materials, but it would be remiss not to touch on other aspects. Returning to Andrews' (1968) list, there is one glaring inclusion, which deals with an application of load in a manner that does not directly relate to the mechanical properties described previously: wear.

There are two obvious examples of wear that can be described to help us define what we are talking about. Firstly, clothing is particularly prone to wear, even if it has not been fashionably distressed. Once again popular, the concept of putting patches on elbows was considered laughable at one point, but came about because otherwise perfectly serviceable jackets were worn out at the elbows, when the arms inside were in contact with, typically tables. This kind of wear is also seen on trousers at the knee. Again, this is the most likely part of the piece of clothing to become worn, as the clothing becomes stretched over this part of the limb, and hence is under tension when it comes into contact with another surface, such as furniture or the ground. The second example to consider is that of the car tyre. There is, of course, a legal requirement in many countries for the tread cut into the tyre to be a certain depth, and this is part of the annual health check of a car, as well as the responsibility of the motorist to carry out spot checks. The road-handling performance of the car is dependent on the condition of the tyre. Tyres become worn through normal driving, although it is worth noting that power-steering can lead to greater wear in certain locations: it is good practice to allow the car to move slightly whilst steering to limit this kind of damage.

7.4 Creep

Creep can be considered as a form of fatigue, in as much as long-term applied load gives rise to changes in the microstructure that lead to eventual failure, which occurs at a load significantly lower than the yield or ultimate tensile strength of the material. However, the applied load in this instance is static, rather than dynamic, and frequently the load in question is self-load. The classic example in this context would be that of the lead used to protect church roofs. In the past the lead would be removed approximately every hundred years, melted, recast, and reapplied for another century. Over this time, the weight of the lead would cause the sheet to stretch, eventually leading to cracking of the sheet, and the potential for water ingress. Mathematically, creep is defined as follows:

$$\frac{d\varepsilon}{dt} = \frac{C\sigma^m}{d^b} e^{\frac{-Q}{kT}} \tag{7.8}$$

where ε is the creep strain, C is a material constant related to a particular creep mechanism, m and b are exponents dependent on the creep mechanism, Q is the activation energy of the creep mechanism, σ is the applied stress, d is the grain size of the material, k is Boltzmann's constant, and T is the absolute temperature. This equation can be modified based on the creep mechanism that controls the failure, such as diffusion of elements dissolved in the matrix, or the movement of dislocations.

7.5 A Note on Corrosion

Corrosion is, fundamentally, the chemical reaction of the material of interest with chemicals arising from the environment. Traditionally, corrosion was the province of metals, and we can easily distinguish between corrosion and the 'rot' observed in, e.g., wood, because rot is essentially a biological function where living organisms eat biological materials (the macromolecules that form the wood). Corrosion, meanwhile is non-biological.

Corrosion can be prevented by the careful selection of materials, and by specifying protective coatings, environmental conditions, and the like. From the perspective of an interested observer, it is interesting to note that archaeological relics can survive for significant periods of time in what might be considered unpromising environments, only to fall apart or corrode rapidly when removed to room temperature, pressure, and humidity. From the perspective of a curator such phenomena can be perplexing and worrying; from that of an observant engineer, it should be terrifying. Take for example the case of artefacts recovered with the excavation of the Mary Rose. The Mary Rose sank in 1545 and spent the next four centuries on the seabed of the Solent. Significant portions of the wreck rotted away, but a meaningful amount was protected by deposits of sediment over time. Personal possessions were recovered that provide insight not only to life on board a Tudor warship, but also to Tudor life more generally.

The Mary Rose Museum in Portsmouth, UK, is well worth a visit: the efforts that have been undertaken to present the ship and the lives of sailors on board at the time of its sinking, not to mention its 33 years of service up to this point, is fascinating, and an exemplar of what such a museum should be.

The curators and conservators have gone to great pains to protect the remains of the vessel: originally raised in 1982, it has taken much of the intervening four decades to preserve the timbers, drying them out by displacing water with propylene glycol, which prevents the wood from becoming too dry and brittle. Today, the remains can be seen close too, and the ship provides an intimate setting for history that can otherwise be somewhat dry and dusty.

But what of understanding failure? The conservators have not only had to deal with preventing wood from degrading but also the unexpected spalling of corroded material from the surface of recovered cannonballs. Advanced characterisation techniques utilising the Dimond Light Source were able to determine which corrosion phases formed during conservation and hence the best route to preventing the situation occurring again (Simon et al., 2018).

Corrosion is also awkward because the same fundamental corrosion processes can lead to different results. For example, Jesson et al. (2013) were able to show that graphitisation, a form of corrosion unique to cast irons, in which the graphite flake within the structure is broken down and back-filled with corrosion products in a manner akin to fossilisation, can occur over a large area leading to a gradual loss of section, or can penetrate into the bulk, effectively creating cracks that act as stress concentrations.

A final thought on corrosion, is that it is not all bad. Corrosion can, surprisingly, be used to protect. When controlled, corrosion processes can be used to create protective coatings. In a sense, this is exactly what occurs with stainless steel, where protective oxides are formed thanks to alloying additions to the steel. In the case of copper, a protective patina can be created through a mix of various copper salts being allowed to form at the surface (see, e.g., Rosales et al., 1999).

7.6 Summary

In previous chapters we have considered the mechanical behaviour of materials and structures on the basis of the microstructure engendered through the manufacture of the component of interest, and on the defects that arise during manufacture. Here, the question of the accumulation of a lifetime of relatively minor degradation processes has been considered, and the impact of these on the ability of the material or structure to withstand operational stresses. These are the kind of issues that lead to structures that have been fine up until now, 'suddenly' failing. The failure wasn't sudden: it may have been approaching instantaneous when the catastrophic event occurred, but the failure has been a lifetime in the making. Understanding the specifics of degradation processes that apply to specific materials and situations can help prevent or at least delay the catastrophic failure, by targeting appropriate maintenance, and predicting when replacement should occur. Monitoring of structures, particularly expensive structures, can also help, especially with extending the life of the structure not through remediation *per se*, but by understanding whether the rate of degradation has occurred as expected, or more slowly. It can also help in preventing catastrophe by determining that the degradation has occurred in an unexpected way.

References

Andrews, E.H., 1968. *Fracture in Polymers*, Oliver and Boyd.

Jesson, D.A., Mohebbi, H., Farrow, J., Mulheron, M.J. and Smith, P.A., 2013. On the condition assessment of cast iron trunk main: The effect of microstructure and in-service graphitisation on mechanical properties in flexure. *Materials Science and Engineering: A*, 576, pp.192–201.

Murakami, Y. and Miller, K.J., 2005. What is fatigue damage? A view point from the observation of low cycle fatigue process. *International Journal of Fatigue*, 27(8), pp.991–1005.

Rosales, B., Vera, R. and Moriena, G., 1999. Evaluation of the protective properties of natural and artificial patinas on copper. Part I. Patinas formed by immersion. *Corrosion Science*, 41(4), pp.625–651

Simon, H., Cibin, G., Robbins, P., Day, S., Tang, C., Freestone, I. and Schofield, E., 2018. A Synchrotron-based study of the Mary Rose iron cannonballs. *Angewandte Chemie International Edition*, 57(25), pp.7390–7395.

Sonsino, C.M., 2007. "Fatigue testing under variable amplitude loading." *International Journal of Fatigue*, 29(6), pp.1080–1089.

8

Testing Modes and Application of Load

The mode by which the inevitable comes to pass is effort.

– Oliver Wendell Holmes, Jr.

8.1 Introduction

When designing a structure that must exist in the real world, it is necessary to think about the kinds of loads that will be applied and how these will be accommodated. For example, Gordon (1978) illustrates the importance of pre-tension with the nature of trees to grow wood that is naturally prestressed, so that when the tree is placed in bending by a strong wind it is able to bend further than if it were not prestressed. In part this comes from the nature of heartwood and sapwood, with the heartwood being stronger in compression and the sapwood in compression, which is why traditional long-bows are made from a single stave of wood which has both heart and sapwood, so that when the bow is bent, the heartwood is in compression and the sapwood is in tension. It is also interesting to see the effect of the wind on tree growth. Where a tree is subjected to a single predominant wind, then it will tend to grow in a lopsided manner to compensate for this loading, which is effectively only acting in one direction. Where a tree is subjected to winds from many directions, then it will grow in a more uniform manner.

Without getting too focused on specifics, there are a range of loading conditions that may need to be considered:

- Loading from self-weight or from earth if a buried asset such as a tunnel or pipeline.
- Intermittent loading arising from the wind, temperature (causing contraction or expansion), or added weight due to rain, snow, ice.
- Impact, such as a bird strike.
- Dynamic, such as traffic loading.
- Internal, such as pressure transient in a fluid being carried in a pipe.

DOI: 10.1201/9780367822347-8

Taken individually, many of these loads can be considered in the long term and their impact on fatigue performance, unless they exceed some critical level. However, what happens when they are considered together? Some forces will act together, whilst others will essentially cancel one another out. Rajani and coworkers have considered this problem in the context of water pipes (see, e.g., Rajani and Makar, 2000), capturing the full range of issues to be considered and hence developing a load case for pipes in service. Fahimi et al. (2016) built on this approach by considering the impact of defects present in a section of pipe, and the varying severity defects will have depending on their location and orientation to principal stresses.

In considering the characterisation of materials however, we need to think about the tests being undertaken and the impact that this might have on the measurements that we take. We also need to consider the difficulty of the test that we are about to undertake.

Take a piece of a material and consider the orthogonal axes (Figure 8.1a). The simplest test that can be undertaken is in uniaxial loading (Figure 8.1b). However, how representative is this of reality, where more complex stress states may occur? What happens if more than one orthogonal stress is applied at once (Figure 8.1c,d)? These constraints will affect the ability of the material to deform plastically and accommodate the forces applied, which will in turn impact on the failure mechanism of the material.

8.2 Primary Forms of Loading

Fundamentally, there are four primary forms of loading, being uniaxial loading, bending, shear, and torsion. All tests will reduce to one of these

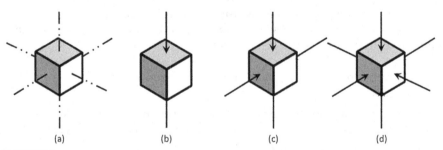

FIGURE 8.1

Consider a cube of material: (a) three orthogonal directions are defined and load may be applied in (b) one, (c) two, or (d) three directions at once.

forms of loading, although sometimes more than one will be used at once; tension-torsion is probably the most common form of compound test. Arguably, shear is a subset based on a specific geometry type combined with a form of loading.

Uniaxial Loading

In some respects, uniaxial loading is the simplest form of testing that is available for mechanical characterisation of materials and structures and it is the basis of much that is used to define mechanical properties and performance. Uniaxial loading encompasses both tension and compression however, and the failures observed through each of these behaviours are different. The failure mode observed is also dependent on the structure of the material. Uniaxial testing forms the basis of most material properties characterisation.

Bending

The simplest form of non-uniaxial loading is bending. This will occur, to some extent, due to the self-weight of the structure, although this will typically be small unless the structure is very large, and/or slender. So, for example, a long hair placed on two supports will tend to sag under its own weight, as will a filament of wire of the same thickness. If we reduce the distance between the supports, the hair will sag less and less perceptibly. However, when it comes to the design process, it is necessary to specify a maximum amount of displacement that can be allowed, and this, together with the overall load to be carried, will define the amount of material required for a beam or similar to cross a known width without exceeding the maximum permitted displacement.

The load to be carried may be localised or distributed. For the latter, think of a lintel above a door or window. Bridges can be more complicated still, depending on the total traffic that they can carry and how much this varies over time. In determining mechanical properties, a sample is usually placed into three- or four-point bending (Figure 8.1). In bending, a material is subjected to both tensile and compressive forces (Figure 8.1b). The maximum strain that the beam sees occurs at the outermost elements, hence I-beams, box sections, and sandwich panels are designed to place the majority of material at the place where the greatest strain is observed. As there is less of a need for materials in the centre of the section, this material can be reduced without compromising the integrity of the structure, producing a beam or panel which is relatively light, thus reducing the potential for deflection due to self-weight. To put it another way, we have increased the specific stiffness (Figure 8.2).

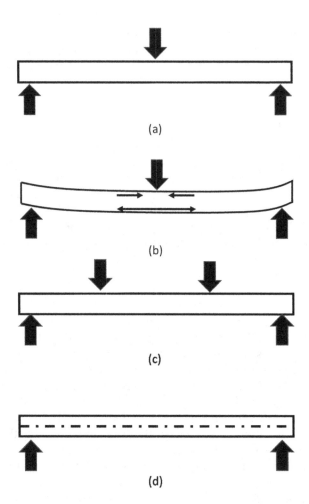

FIGURE 8.2
Illustration of the tensile and compressive forces placed on a structural member in bending.
(a) A sample placed in three-point bending, (b) the specimen in (a) sees compressive force on
the top face and tensile force at the bottom, because the curvature engendered in the specimen
leads to the bottom face becoming stretched and the top face compressed. (c) A sample placed
in four-point bending. (d) A sample in bending demonstrates a neutral axis, nominally through
the centre of the specimen, where the material is neither in compression or tension. The neutral
axis will move towards the stiffer face, which may occur if, for example, the material is stiffer in
either tension or compression, or if the material is in any way damaged.

For example, simple bending theory, which can be applied to both three-
and four-point flexural tests, can be reduced to:

$$\frac{M}{I} = \frac{\sigma}{y} = \frac{E}{R} \qquad (8.1)$$

where M is the moment,[1] I is the second moment of area,[2] σ is the stress, y is the distance from the point of loading to the neutral axis, E is the stiffness, and R is the radius of curvature. This equation is accurate, where R is large, i.e., the overall deflection of the specimen is small, but will break down when the defections are large, and correction factors will need to be applied.

It can be seen, therefore, that we can adapt this to a particular set of circumstances by introducing specific values for the variables in the case. For example, when designing a strut, or some other structure, known strength data for a material under consideration can be supplied, and the second moment of area for the design, and hence we can determine the maximum expected deflection of a given applied load. In the case of a physical test, perhaps of a new material, or a proof-test of a production run, test data can be collected (load, displacement) in order to determine the stress, and ultimately the strength of the material. The behaviour of the material itself is possibly not of interest.

However, there was a situation in a particular industry, where very capable engineers were not able to determine the cause of failure of a series of components. Failures were not unexpected, in part because the components in question had gone well beyond the design life-time. Corrosion played an important part in the lifetime degradation of performance. Standard practice was to determine the extent of corrosion, and hence to carry out a 'loss of section' calculation, which is to say, a calculation of the residual strength at the time of failure, based on the remaining uncorroded material, and to check that this was congruent with the known operating conditions at the time of failure. For the most part this had been shown to be a sensible strategy, but there were an increasing number of failures that did not fit this model. The failures were inexplicable using this approach, and there was no satisfactory explanation which could suggest the factors that were contributing to the failure: given the residual 'good material', these components should not have failed. And yet here were a number of demonstrably broken components. Under such circumstances it is best to follow the counsel of Bertrand Russell:

> *In all affairs it's a healthy thing now and then to hang a question mark*
> *on the things you have long taken for granted.*

External advice was sought, and the sector was introduced to fracture mechanics: cracks lead to stress concentrations at the crack tip and in this

---

[1] A moment, also called a moment of force, acts to make an object rotate around a fixed point. The size of the moment is a function of the distance from the pivot point, and the force applied. A perfect example of this is the use of an allen key. When undoing a hex-bolt, you get a larger moment if you use the longer arm for leverage. When the bolt is looser, it is quicker to undo it when the shorter arm is used to turn the key.

[2] The second moment of area, sometimes called the moment of inertia, is a geometric function which addresses the impact of the cross-section on various other functions. Perhaps the simplest is a rectangular cross section, where $I = \dfrac{bd^3}{12}$.

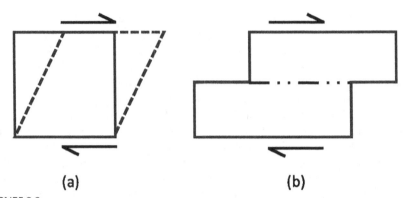

FIGURE 8.3
Illustrating the action of shear: (a) deformation due to a shear load and (b) a sample failing in shear.

instance some forms of corrosion can act in a crack-like manner. These were shown to be the cause of the unexplained catastrophic failures, but it required a re-evaluation of the material and its behaviour when degraded, to provide an answer when the conventional wisdom of the industry had been shown to lack an answer. This was considered in greater depth in Chapter 5.

The point is that not all materials behave in the same way. In the context of materials performance there are a number of factors which affect the mechanical properties of a material, which is why metals and ceramics, for example, (tend to) behave in such different ways. But even if we focus on only one material, various factors can affect performance. This will be explored further in the context of strain-rate dependence in Chapter 10.

Shear

Shear occurs when a specimen is loaded normal to the face that it is acting on, rather than parallel, with one face perpendicular to the applied load fixed in place (Figure 8.3a). In uniaxial tension or compression we take the sample and pull (or push) along the main axis, and cause the specimen to elongate (or compress). With shear, we take the same specimen, hold it in the same way at the bottom, but then push at the top, leading to deformation, with final failure occurring as seen in Figure 8.3b.

Torsion

Like shear, torsion occurs when force is applied in a non-axial manner (Figure 8.4), such that deformation of the material occurs. Torsion is a fundamental aspect of many engineering processes, including, for example, the transmission of force from an engine to a wheel-set using a drive shaft.

Another example familiar to many is that of a screw being driven into a bulk material such as a masonry wall or a wooden fence. The screw is rotated, and due to the force placed behind and the pitch of the thread, the screw is driven in, thereafter to hold something in place. However, if the screw is unable to move and we continue to apply force, the screw will fail through torsional shear.

$$T = \frac{J_T}{r}\tau = \frac{J_T}{l}G\varphi \tag{8.2}$$

where T is the applied torque or moment of torsion, with units of Nm, τ is the maximum shear stress at the outer surface, J_T is the torsion constant for the section, r is the distance from the rotational axis to the outermost surface point, which will be the radius for a circular cross-section, but could be a corner for a square cross-section or some other feature for an irregular object, l is the length of the object over which the torque is being applied, φ is the angle of twist in radians, and G is the shear modulus.

8.3 Plane Stress and Plane Strain

Having suggested that there are some complexities to be associated with the behaviour of materials, at this point we should recall the introduction of Poisson's ratio in Chapter 5, and indeed the issues surrounding cracks in Chapter 6. In the case of Poisson's contractions, this effectively represents a reaction to the force applied, whilst in terms of cracks we would like to control the manner in which they grow, without interfering overly with the mechanisms that we are trying to observe. Here, plane stress and plane strain can be used to help us, although it is always necessary to consider what the impact of simplification will have on translating experimental results to a real-world scenario.

Plane Stress

Recalling the sample of material in Figure 8.1, and identifying the orthogonal directions (Figure 8.5a), we can undertake mathematical operations to calculate the various stresses we are interested in. However, we can simplify these operations by considering a situation of plane stress, whereby if we define a plane xy, the through-thickness direction, z, is infinitely thin (Figure 8.5b). Hence, there will be no stress, other than that in the x and y directions. In practice, the materials that we consider are not going to be infinitely thin, but

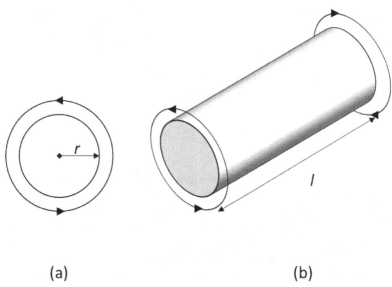

FIGURE 8.4
Effect of torsion on a cylindrical sample.

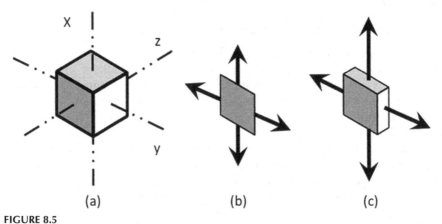

FIGURE 8.5
Recalling Figure 8.1, we can (a) identify specific orthogonal directions. In the case of an infinitely thin xy plane (b), there can be no through-thickness stress in the z direction. For a real material, there is a finite thickness (c) for which this assumption can be considered to be true.

this approach can be considered valid if the through-thickness is sufficiently thin. In the context of a thin-walled pressure vessel, this is usually taken to be:

$$\frac{r}{t} < 10 \qquad\qquad (8.3a)$$

or

$$t < \frac{r}{10} \qquad (8.3b)$$

In other words, if the radius of a tube were 10 mm, then the wall thickness would need to be less than 1 mm for this assumption to be valid. Happily, this is frequently easy to achieve, although it should be noted that the microstructure of the material may become important, as indeed will the processing route. Some microstructural features may be suppressed or take unusual forms if the thickness of the material is too small, and this may affect the properties of the material. As we increase the size of the component, different microstructures might be observed, and hence a different strength will be seen in a smaller component compared with a bigger one, unless the processing route is controlled carefully.

Plane Strain

By contrast, in the case of plane strain, strain is assumed to be zero in the through-thickness direction, when the specimen is sufficiently thick that it constrains the material from moving when the sample is loaded. This sort of condition will occur in long billets of material, and in structures with thick walls, a prime example being hydraulic dams. A stress is generated in the through-thickness direction (Figure 8.6) which is defined as:

$$\sigma_z \approx \upsilon \left(\sigma_x + \sigma_y \right) \qquad (8.4)$$

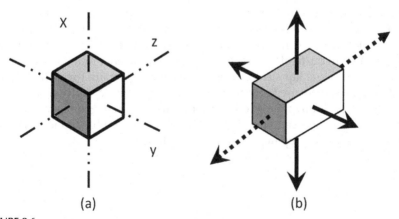

(a) (b)

FIGURE 8.6
In plane strain, a stress is generated in the through-thickness direction (as indicated by the dashed arrows) due to constraint by the bulk when the material is loaded in other directions.

The key thing to make note of, however, is that around crack tips, two very different behaviours are observed. In the case of plane stress, there is significant plastic yielding and the maximum fracture toughness is achieved. Where a specimen is in plane strain, a triaxial stress state is found at the crack tip such that plastic yielding is constrained, and hence the minimum fracture toughness is found. There is no point of inflexion between the two conditions, but rather a region of mixed-mode behaviour between the two extremes. Recalling Figure 8.4b and the associated discussion, it is also worth noting that, of course, an infinitely thin plate will not have any volume to plastically deform and so there will be a drop off in fracture toughness beyond a certain thickness.

8.5 Modes of Testing for Investigating Fracture

In terms of testing samples containing cracks there are three modes of testing to be considered (Figure 8.7).

It will be noted that Mode I is essentially uniaxial loading, whilst Modes II and III are both forms of shear. Mixed modes of failure are possible, although testing such specimens is non-trivial, and will frequently require sophisticated jigs to manage the loading of the specimen.

8.6 Summary

Testing specimens to determine material properties can be very straightforward. It can also be very complicated, with the difference frequently arising from the nature of the material being tested and the requirements to constrain the specimen in particular ways to prevent certain behaviours occurring. For example, the single lap shear test is fairly easy to prepare for, although care must be taken to ensure that the individual specimens are properly aligned. However, there is an inherent danger with this specimen geometry that could lead to out-of-plane bending of the specimen. The double lap joint is more time consuming to prepare, but more stable to test.

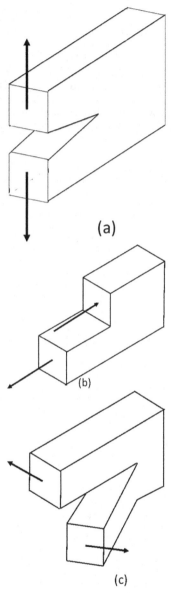

FIGURE 8.7
Modes of testing: (a) I – opening, or tensile, (b) II – sliding, or in-plane shear, and (c) III – tearing, or antiplane shear.

References

Fahimi, A., Evans, T.S., Farrow, J., Jesson, D.A., Mulheron, M.J. and Smith, P.A., 2016. On the residual strength of aging cast iron trunk mains: Physically-based models for asset failure. *Materials Science and Engineering: A*, 663, pp.204–212.

Gordon, J.E. 1978. *Structures or Why things don't fall down*, pp.149–167, Penguin.

Rajani, B. and Makar, J., 2000. A methodology to estimate remaining service life of grey cast iron water mains. *Canadian Journal of Civil Engineering*, 27(6), pp.1259–1272.

9

How to Measure Strain

I often say that when you can measure what you are speaking about, and express it in numbers, you know something about it; but when you cannot measure it, when you cannot express it in numbers, your knowledge is of a meagre and unsatisfactory kind; it may be the beginning of knowledge, but you have scarcely, in your thoughts, advanced to the stage of science, whatever the matter may be.

– William Thomson, Lord Kelvin
"Electrical Units of Measurement" in
Popular Lectures (1883) Vol. I, p. 73

9.1 Introduction

Strain is one of the fundamental measurements that can be made when undertaking mechanical characterisation. On applying a load, we begin to change the specimen that we are testing. Initially this change will be all but impossible to identify, but whether we are pulling, pushing, bending, or twisting, we are making the atoms move relative to each other. This was discussed in detail in Chapter 4. From that discussion, it was shown that we can identify a property that represents the change in the specimen, which we call strain. As will be recalled, strain is the change in original length, divided by the original length. The original length is relatively easy to determine, and to determine accurately – although it is important to beware of spurious precision in quoting the dimensions of a component. For example, if you have a rule marked off in millimetres, and make multiple measurements, you cannot provide an average measurement to the nearest 100 microns.

When the test is underway though, how are we to determine the change in length? In a very basic test, we might use the same millimetre marked rule used to determine the original length, but that is rather unsatisfactory, for a number of reasons, as a moment's thought will show. At the very least, there is the question of keeping the rule steady against the specimen, which might affect the specimen as it is being tested, not to mention the potential safety problems, especially if you are taking your test to failure. A common methodology in civil engineering is to use a DEMEC (demountable mechanical)

DOI: 10.1201/9780367822347-9

strain gauge, developed by Morice and Base (1953). Today there are numerous suppliers of the equipment for this technique, but the basic principle is that small discs of, typically, mild steel[1] which have a central dimple, are bonded to the surface of interest. A bar, usually of invar[2] (although sometimes this is simply used for calibration purposes), with two conically tipped pointers and a dial gauge is placed with the tips of the pointers fitting into the dimples. One pointer is fixed, and the other is on a knife-edge pivot, causing the dial gauge to register a change in position, which is typically expressed as a strain.[3] The original paper suggests that a skilled operative can conduct some 200 measurements an hour, with an accuracy of $\sim \pm 5 \times 10^{-6}$. The accuracy can be improved upon under ideal laboratory conditions. Like many simple concepts, it is easy to understand, easy to learn the basics, but does require some practice to do it well and develop the skill and precision to achieve the rate of testing and accuracy stated in the paper.

This is just one way of determining strain: it has its advantages and disadvantages. This chapter will explore some of the most commonly used methods to determine strain in and out of the laboratory environment, ranging from simple measurements to the latest computer-vision-based approaches.

9.2 Crosshead Displacement and the Problem of Machine Compliance

When undertaking a test in the laboratory, whether it is a uniaxial test (i.e. pure tension or pure compression) or a more complicated test geometry (such as an Iosipescu[4] shear test) the test will be undertaken in a test frame. There are three main classes of frame: the 'universal' test frame (typically screw-driven), the servo-hydraulic system, and the relatively recent addition, the linear motor device. Whatever the engineering basis for the application of load, there are four key features to consider (Figure 9.1): clearly

---

[1] Sometimes stainless steel or other materials with good weathering characteristics are used if the installation is to be used for a longer-term programme of strain monitoring.

[2] An alloy of iron and nickel, invar is frequently used in scientific instruments and mechanisms thanks to a co-efficient of thermal expansion that is virtually zero.

[3] With respect to the point made about health and safety, there are two modes in which this kind of measurement is made. Firstly, the system is under natural load and the structure being assessed is undergoing routine and long-term asset management in order to ensure that all is well. Secondly, whilst the measurements may be made on a sample undergoing active loading, this will not be a quasi-static test, but rather increases in load will not continue until the measurement is made.

[4] Frequently misspelled as 'Iosepescu'; the methodology for this test is found in ASTM D5379, originally published in 1992 and last revised in 2012.

there has to be some kind of frame to hold the test, there must be some means of measuring the load applied to the specimen to be tested, there must be some way of holding the specimen, and there must be a moving part in order that the specimen can be loaded.[5] In the case of the universal test frame, it is the crosshead that moves. Servo-hydraulic and linear motor frames have a slightly different design, whereby a piston is driven up and down either by pumped fluid, usually oil, or by an electric motor. One set of jaws, or some other attachment which connects with the specimen, are directly attached to the piston. Even so, similar principles are in play, and so we will consider the case of crosshead displacement and treat all three frames the same knowing that there are some fundamental differences in operation but that the mechanical problem that we are currently interested in is the same whatever test frame we actually use for the test.

Crosshead displacement is perhaps the simplest method available for determining strain, especially as most commercial test frames will record this information automatically when a test is underway. When this happens, the specimen will resist the deformation that is caused by the movement of the crosshead (or the piston), and it is this resistance that is measured by the loadcell. The problem, however, is that no test frame is infinitely stiff. The consequence may be expressed in the same terms as Newton's Third Law of motion: for every action, there is an equal and opposite reaction. To put it another way, if testing an infinitely stiff specimen, the test frame would fail in the attempt of trying to test it (which, incidentally, is why it should be standard operating procedure to ensure that physical trip limits are set on the test frame as well as virtual ones on the control software). Every component of the test, as detailed in Figure 9.1, contributes something to the stiffness observed in the load–displacement output of the test. Once plastic deformation, or some other indication of ongoing failure,[6] is underway, the effect of the test frame can be ignored. However, whilst mechanisms of failure are important, it is the behaviour of the material prior to this permanent change that is typically most important to designers.

Compliance, C, can, in some respects, be considered as the inverse of stiffness; it is the ratio of displacement, δ, and load, P, i.e.:

---

[5] In some circumstances the specimen changes due to some external stimulus and hence causes the linkage to the load cell to move, and the load measured is a consequence of this movement. Jesson et al. (2010) describe a test on exhumed cast iron water pipes, which included tests with the specimen – a 4″/100 mm internal diameter pipe from a distribution main – fitted with special end caps which allowed the specimen to be held in place, and which also allowed the pipe to be filled with water whilst under test, and for the water to circulate. The water was passed through a chiller and the temperature of the water was varied. As a consequence of thermal contraction, the load induced as the pipe cooled could be measured.

[6] See, for example, the behaviour of fibre-reinforced cements and concretes, where the slip of fibres is a non-reversible form of damage but does not conform to the definition of plastic failure given in Chapter 4.

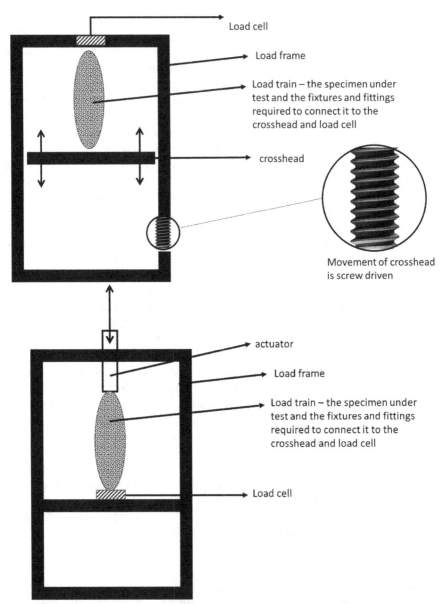

FIGURE 9.1
Schematic of the principal components of a test frame: (a) a universal test frame, where the entire crosshead is moved and (b) a test frame where only the actuator moves, driven either by a servo-hydraulic or electro-mechanical system.

$$C = \delta/P \tag{9.1}$$

with an SI unit of m/N, although it is more likely to be expressed in terms of mm/N. Considering Figure 9.1, every part of the load train, together with the overall frame, will contribute to the total compliance, C_T, of a test set up consisting of compliance 'elements', from the test specimen, C_S, and from different parts of the load train including the load cell, grips, adapters, and the frame itself ($C_1 + C_2 + C_3 \dots$):

$$C_T = \Sigma(C_S + C_1 + C_2 + C_3 \dots) \tag{9.2}$$

Each element can be considered if it is thought to be necessary. The perspicacious will realise that (a) such a calculation will need to be carried out for any given test geometry (because the load train will be different for a tensile test specimen[7] and one in compression for example) and (b) the focus of the calculation can be changed:

$$C_T = C_S + \Sigma(C_1 + C_2 + C_3 \dots) = C_S + C_M \tag{9.3}$$

where C_M is the machine compliance (if we take a broad view and include the full load train with this), which is to say that in some respects the actual test geometry is of relatively little significance. It is possible to determine the compliance of the test frame by taking a very stiff sample, typically a large piece of metal, and applying a load to it. The applied load should be no more than a few percent of the load cell capacity.[8] Under such circumstances the specimen will be loaded but will not be deformed, and any crosshead displacement registered will therefore be due to the machine compliance. Therefore, when $C_S \rightarrow 0$:

$$C_T = C_M \tag{9.4}$$

When undertaking actual tests using this setup, the load–displacement data can then be corrected, firstly by subtracting the machine compliance:

$$C_S = C_T - C_M \tag{9.5}$$

And then by expanding and rearranging:

$$\delta_S/P_S = \delta_T/P_T - \delta_M/P_M \tag{9.6}$$

---

[7] And indeed, even tensile specimens of different thicknesses might have an effect, because different jaw faces will need to be used.

[8] In some instances this will be the same as the frame capacity, but some frames are designed to be used with a range of load cells, and some of these can be considerably smaller than the frame. Applying 1 kN to a 100-kN load frame with a full-capacity load cell is perfectly reasonable, but when using a 10-N load cell, it is not.

$$\delta_S = P_S(\delta_T/P_T - \delta_M/P_M) \tag{9.7}$$

But the load that is applied to achieve the total displacement is the same load seen by any part of the load train, including the specimen that we are interested in. Therefore:

$$P_S = P_T = P_M \tag{9.8}$$

Which means that equation (9.7) can be simplified to:

$$\delta_S = \delta_T - \delta_M \tag{9.9}$$

The theory is impeccable, and the practitioner should always attempt the process at least once to convince themselves that it does indeed work, whilst at the same time remaining the least accurate method of ultimately determining strain. The strain output from crosshead measurements is inherently flawed due to machine compliance issues, and whilst some corrections can be applied, corrections that will be unique to a particular test frame, the final result is still likely to be overly compliant and give a spurious measure of the stiffness of the specimen.

9.3 Strain Gauges

This chapter begins with a quote from William Thomson, better known as Lord Kelvin, the 19th-century physicist who made his mark in multiple areas; his areas of work included thermodynamics, electricity, maritime navigation, and telegraphy. As well as being a mathematician and physicist, he was an inventor and engineer. One of his biggest successes, certainly from the perspective of someone interested in mechanical characterisation, is the invention of the strain gauge (e.g., Thomson, 1883).

The fundamental principle of a strain gauge is that a wire of a certain diameter has a certain resistance, which can be measured. If the wire is stretched, the resistance of the wire will change, and this deviation can also be measured. This is not very useful in the case of some metals, which require a significant elongation before there is detectable change in the resistance, one must therefore stretch a long piece of wire by a considerable amount. However, some metals are more appropriate: a thin, relatively short piece of wire can be used. If, then, a piece of wire is curved back on itself several times, a relatively small gauge can be made, which is precise to ~ ±10 µε (i.e. 1×10^{-5} ε or 0.001 %ε). For ease of use the wire is usually encapsulated in a polymer film, and part of the art of selecting the appropriate strain gauge (and indeed designing them in the first place), is in choosing the right polymer sheath/wire combination for the job at hand.

There are several manufacturers of strain gauges, and it is possible to purchase strain gauges for almost any conceivable situation: material to be tested, strains (single or multiple, direct or shear), operational requirements (environmental conditions, longevity of test) can all be taken into consideration, for example. Multiaxial strain gauges, typically referred to as rosettes, are available, which whilst typically more expensive than just using three separate gauges, have two distinct advantages. Firstly, the strain gauge collects information from a single point of interest, and there is no need to compromise in the placement by trying to fit three different gauges into the available space, and secondly, the total work in installing the gauges is much decreased. This second factor should not be overlooked when a test regime is being planned (although the extra expense means that this is not a job to delegate to a novice).

There are several key factors to consider when selecting strain gauges. The fundamental one is of course the nature of the material itself. Is the material likely to become internally heated as a result of the test being undertaken? (This is a particular issue with fatigue tests undertaken at relatively high frequency.) Does the material have a heterogeneous microstructure, such as that found in composites and concretes? (The standard for concrete is to use a gauge that is five times as long as the diameter of the aggregate. This can be adapted to a good rule of thumb for composites, which is to say that ideally one is looking for a strain gauge that is five times the length of the unit cell of the fibre architecture.) How long will the test last? Will the specimen be retested at a later date? What are the operating conditions? Is this a test in the laboratory or in the field? Any reputable supplier maintains a catalogue of available gauges and the circumstances in which they should be used, and it is well worth taking the time to contact the supplier and discuss the test requirements. Where possible, reputable suppliers will be happy to visit your facility, or for the researcher/technician to visit them, and to deliver training. Even experienced users seek the advice of experts when dealing with unfamiliar strain gauges for the first time, such as those used to monitor crack growth (which have strands that break as the crack grows) and those for measuring shear during the Iosipescu test. Most suppliers also maintain an internet-based archive of videos and documents to help teach the basics of installing strain gauges. Such material takes in the basic considerations of specimen preparation in order to achieve an optimal installation. Adhesive type is also considered, and again this can affect the test, and the longevity of an installation. One does not want to end up measuring the strain in the adhesive instead of the specimen, because the wrong kind of glue has been used. Typically, a glue that is neither too ductile nor too brittle is required, and superglue can be used, although a primer or adhesion promoter may be required.

Recalling from Section 9.1 that a skilled operator can undertake of the order of 200 measurements an hour using the DEMEC system, a well set up datalogger, collecting information from a well installed strain gauge can collect several times this number of measurements every second, and therefore the question becomes one of determining a meaningful sampling rate in order that the glut of data does not become overwhelming.

The average practitioner will probably only need to keep a handful of types of strain gauge in mind for the tests that they will typically undertake. It must be remembered though that the key thing that sets human beings apart from most other animals is imagination. Hence, if you can imagine a test, there is probably a strain gauge already available for it and, if there isn't, with modern technology it would not be too hard to make a custom gauge for a particular experiment.

However, a careful read of the catalogue[9] is also a sensible idea, and a discussion with your regular supplier can pay dividends. There are gauges which are tuned to particular materials; there are gauges that are more suited to fatigue tests; there are all manner of special gauges for determining strains under odd loading conditions and in strange places.

One final note, born out of experiences with gauges to measure crack growth, in the majority of situations, all things will be well. However, very occasionally they won't: it is entirely possible for spurious readings to be recorded during a test, and for these to cause great consternation. In the case of a crack-measuring gauge, the gauge is designed to determine the growth of a crack during cyclic fatigue. It does this in a stepwise fashion: the gauge is glued to the specimen where the crack is due to grow. Strictly speaking this is not a strain gauge in the sense that we have been discussing, but it is used for mechanical characterisation, and does operate on the same basic principles as other wire-based gauges. In this instance though, the gauge forms a circuit of wires in parallel and in pristine condition all the wires carry the current that is passed through the gauge. As the crack grows, each wire is broken when the crack tip crosses it. However, the broken wires of the gauge can end up touching and hence conducting current. Careful observation is required during the test, and some experience is helpful when reviewing the test data. This is perhaps the most extreme thing that can occur, but it is entirely possible to install the gauge upside down (i.e. with the metal contacts against the specimen), use the wrong gauge for the experimental set up, or even for it to be not quite aligned properly with the loading direction. It is important to review the information collected: the practitioner must assume that anyone reviewing their data will be judging it, and that it will need to be defended.[10]

---

[9] As a certain generation of lecturers are keen to point out, a month in the laboratory can save an hour in the library.

[10] Although it is often said that the modeller always believes their model, and no one else does, whilst a physical researcher never believes their data, when everyone else takes it at face value.

This is good practice in any branch of research, and a good defence is one that is grounded in a thorough knowledge of the material tested and its normal behaviour, the operation of the test, and an understanding of the typical sources of error.

9.4 Mechanical Extensometers

A major manufacturer of mechanical testing equipment suggests that there are fourteen ways of measuring strain. This might even be true, but in the context that they are using it, it is perhaps slightly misleading. They do not mean fourteen different kinds of method, but rather are talking, predominantly, about the different kinds of extensometers that are available for measuring strain, or some associated characteristic, in different situations. And if we are being precise, then they offer fourteen different kinds of extensometers, but there are other kinds as well, not included in their catalogue, some of which are for very specialist applications, such as for use in bore holes. This section focuses on the general situation of 'mechanical' extensometers, which is to say extensometers that are attached to the specimen. Virtual extensometers, i.e., those that make measurements in some non-contacting way, are considered in the next section.

The first mechanical extensometer as we would recognise it today was described by Huston (1879), who describes a piece of apparatus used to determine the extension in a piece of iron, and to compare the elasticity of the test piece with the elongation arising from an applied load, the speed with which the load was applied and the length of time for which it is held.[11]

The most straightforward extensometer is the simple clip-on (Figure 9.2). In this situation stiff elastomeric bands (O-rings tend to be very good for this) are used to hold the extensometer in place, whilst 'knife edges' push into the surface. On surfaces with a low coefficient of friction, such as polished metal or composites which have been made on flat tooling, the knife edges can be prone to slippage, making measurement difficult. In some circumstances, a small blob of hot-melt glue, liquid metal adhesive, or 'plastic padding' can be put on the surface, and a notch placed into the blob prior to solidification/curing, into which the knife edges can be placed to give better purchase.

More complicated extensometers can be used to measure crack opening displacement, or the circumferential strain in a pillar, for example.

---

[11] It is worth seeking the paper out, not just for its novelty value. It provides a useful exercise for the researcher in considering the fundamental issues to be presented when discussing the results of an experiment, whilst at the same time emphasising the importance of a good literature review.

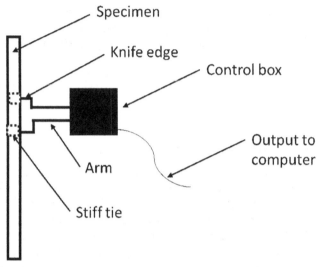

FIGURE 9.2
Schematic of an extemsometer attached to a specimen.

9.5 Virtual Extensometry and Digital Image Correlation

All methods of determining strain have their advantages and disadvantages. There are two key issues with measuring strain directly from the specimen with either strain gauges or an extensometer. The first is that the information that you receive as an output will tend to be highly localised or it will be averaged over a larger distance. Neither situation is very satisfactory when dealing with a material containing defects or in some other way having significant variation at the microstructural level, such as composite materials and concrete. The second, which admittedly is a relatively uncommon problem, is what we might refer to as the Heisenberg issue: we cannot always be sure that in applying the gauge or extensometer that we have not affected the sample in some way. It is incredibly easy for a novice to misapply a soldering iron for example, or for a specimen to be treated in a heavy-handed fashion when ensuring that the bands holding the extensometer in place are sufficiently tight.

Fortunately, there are non-contacting methods. The general feature of these methods is that a camera, or a suite of cameras, is used to monitor markers that are placed on the surface of the sample of interest. At its simplest a single camera can be used to track two markers: this is usually referred to as virtual extensometry, and is relatively easy to carry out. Whilst the specifics vary from system to system, most of the major test frame manufacturers have a bespoke system which can be sold as an accessory with a new test frame, or as an upgrade. Most are also compatible with other systems or can be used in conjunction with custom-built self-reacting frames.

A variant of this is laser extensometry: in some ways this is even simpler than video extensometry, in that you do not need to apply markers to the specimen for the tracking algorithm to follow. Instead, a laser beam is directed onto the specimen and the reflections are captured by a CCD camera and analysed.

Virtual extensometry is like other forms of extensometry in that you can get continuous and live strain data output whilst you are carrying out the test. In some respects, it is more complicated because now there is the additional issue of lighting to consider: even in a well-lit laboratory, additional lighting is usually necessary to ensure that there are no opportunities for visual artefacts to be created, such as shadows or point reflections. That said, the engineering principles behind the technique are straightforward: instead of measuring the change in resistivity of a wire as it becomes stretched, a computer program measures the distance between two markers. If continuous data are not required, those so inclined can achieve a similar effect using an SLR camera and digital image management software, such as ImageJ.[12] Originally developed for the analysis of images of biological origin, such as cell tissue, the program has a great deal of functionality which can be applied in a number of other contexts. One of the main features of the program is the ability to batch process images, and to track features. For example, Boughanem et al. (2015) assessed the orientation of fibres in a reinforced cement sample, analysing hundreds of images per sample, whilst Matthews et al. (2020) used ImageJ to (1) produce scale bars on images; (2) undertake greyscale profile plotting for an analysis line and as a histogram of an image; and (3) stitch together multiple images. The program is open-source software[13] and there are a number of plugins available, some of which are bundled with the main program in 'flavours' available from other sources. As the program is open source, it is relatively easy for the user to develop their own plugins, and there is a wide array of user forums for help with coding issues.

Digital image correlation may be thought of as a form of virtual extensometry in as much that the sample is photographed as the test proceeds. The point made above about lighting remains, if anything becoming more critical. However, unlike other forms of virtual extensometry, significantly more information is captured, although the consequence is that it is rarely feasible to record strain information for the entire duration of the test. Instead, it is more usual to capture data every few seconds, or at significant points in the test. As a consequence, DIC should not be considered as a default alternative to the humble strain gauge or extensometer: there is a trade-off to be considered in terms of the set-up time vs. the (usefulness of) the information collected. Fundamentally, do you need full field strain data?

---

[12] ImageJ is a java-based program developed by the US National Institutes of Health.
[13] It can be downloaded from https://imagej.nih.gov/ij.

Why is so much more data collected? In digital image correlation, the algorithms of a computer program track features on the surface of the specimen, and generate a full field strain map, based on the measurement of full field displacements. This means that you can determine ε_{xx}, ε_{yy}, and τ_{xy} simultaneously, and consequently calculate Poisson's ratio.[14] Most materials do not have a sufficient density of features to be tracked in their natural state, and so it is necessary to apply features. These features are typically referred to as speckles. The distribution of speckles must form unique, random neighbourhoods, that can be used to form 'subsets'. A subset is a small area, a square array of pixels that form a region of the area of interest (AoI) under consideration. The total number of pixels will be a function of the camera used but will typically be a few megapixels.

To begin with, it is useful to think in terms of the 'five' on a die: we have five circular spots within a square (Figure 9.3a). Under load, the relative positions of the spots will change (Figure 9.3b); displacements, and subsequently strains, can be calculated. In theory, it might be supposed that the perfect speckle would be one that would give rise to a 3 px × 3 px subset, a speckle being 1 px (Figure 9.3c). Of course, if we put a couple of these subsets together, with their opposites (swapping the black and white pixels around), it is immediately obvious that there is a problem: the recognition algorithms will become confused because the pattern is so regular – there are no unique subsets (Figure 9.3d). In practice, therefore, we do not want the speckles to be perfectly arranged, but rather they should be randomly positioned within the subset (e.g. Figure 9.3e). Further, current best practice is that five speckles within a subset is insufficient, and that nine speckles are required. These speckles do not have to be perfect circles, and indeed it is better if they are not (Figure 9.3f).

---

[14] The picture on the cover of this book is an example of the output from DIC. The specimen is in four-point bending, and the resulting rainbow effect is a result of the transition from compressive strain at the top, through the neutral axis in the centre, to tensile strain at the bottom. Concentrations of strain can be observed around the diamond cut out from the centre of the specimen. The distribution of strain is, of course, more smoothly linear in reality, but the output is 'binned'. The number, and indeed the size, of bins can be varied as required, depending on the software used. Again, different software packages provide the output in different ways, but generally the scale will automatically update as you progress through the specimen. This can be useful to show the evolution of strain concentration but can give a spurious confidence in the overall magnitude of the strain: early in a test it may appear that there is a significant strain concentration in a particular location, when in fact the bins are separated by only a few hundred microstrain. However, it is usually possible to determine the final maximum tensile and compressive strains and to set the scale for every image to these. It is then possible to track backwards and forwards when strain concentrations emerge, especially where these lead to failure, and perhaps just as importantly, when they don't. An early lesson for the practitioner is in the interpretation of the colour patterns produced by DIC and learning to match these to the physical measurement. There are a lot of pretty patterns that are produced which are essentially meaningless because the whole scale is still in the noise.

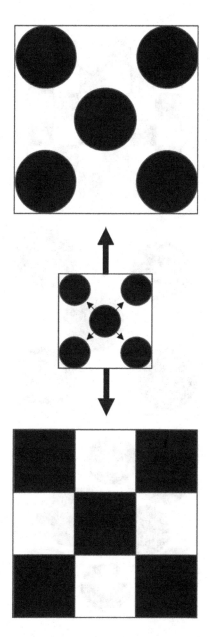

FIGURE 9.3

Understanding the speckle pattern for digital image correlation. A subset, i.e., a unit area being assessed, needs to have a certain number of features. For the purpose of illustration, we begin with five (a), arranged as the five-spot on a die. Under load (b), the relationship between the spots changes, and it is the measuring of these changes that gives rise to a determination of strain in the specimen. If we idealise, this relationship, we give rise to something like (c), but if we consider (d) we will see that the ideal falls short, because we need the pattern to be unique in each subset. Therefore, we want the speckles to be randomly positioned within the subset, (e). In reality, we would like more speckles in the subset, ideally around nine (f).

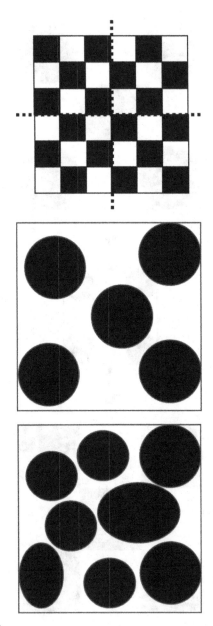

FIGURE 9.3 (Continued)

The methods for applying speckles are numerous and limited only by the ingenuity (or overall knowledge of the literature) of the operator. For some time, the default method for applying speckles was by spray painting the specimen, usually by applying a base coat of white, and then misting black paint over the surface. However, this method has fallen out of favour for two key reasons. Firstly, the technique gives rise to speckles that are usually quite fine, which is to say that the density of speckles is high, but that they are very small and lacking in detail, and with limited contrast to the basecoat. This means that the output is prone to a problem called aliasing, whereby the computer algorithms that track the speckle-features get confused, which has an impact on the accuracy and resolution of the output. The second issue is that it becomes difficult to ensure consistency, even when the speckle pattern is applied by the same operator. Matthews (2020) discusses these points in more detail and has proposed a new approach to optimise the speckle for the application and a new methodology for printing a speckle pattern onto a kind of transfer paper. Whilst not applicable in every situation, the approach is a significant improvement over previous attempts to print speckles onto a vinyl sticker,[15] because the paper has no measurable impact on the performance of the test. The application of the methodology in several other extreme situations gives confidence in its wider use (Matthews, 2020).

There is some debate over the 'perfect' speckle pattern. Jones and Iadicola (2018) have brought together the collected knowledge and experience of the DIC user community, to produce *A Good Practices Guide*. This covers all of the key areas to be considered when undertaking DIC and is essential reading. With respect to speckles however, there is so much that is context specific, and such a variety of advice in the literature, that the *Guide* does not give any specific advice. Conventional wisdom suggests that the pattern produced should be ~50:50 black and white, but Matthews (2020) includes an in-depth study which suggests that in many contexts a ~70:30 balance in favour of the speckle may be preferable. Perhaps more importantly, Matthews (2020) provides a methodology for determining the optimum speckle in a given context in a systematic manner.

A variant of this form of extensometry is called the grid method. Here, instead of using a randomised pattern of speckles, a uniform array of identical points is tracked (see, e.g., Pannier et al., 2006). This technique is considered to be particularly useful when carried out in connection with the virtual fields method, and for providing information for complex stress states. The technique is usefully reviewed by Grédiac et al. (2016). It is limited in that, like 2D-DIC, it does not provide any out-of-plane information.

---

[15] Such as those used for applying company branding to vehicle coachwork.

9.6 Summary

This chapter has introduced several methods for the determination of strain. When dealing with very simple tests, determining strain is relatively easy. It must be remembered that most methods of determining strain deal either with a localised measurement, or one that is an average over a smaller or greater distance. Determining the strain over a large area requires some thought: simply plastering the specimen with strain gauges leads to a complicated test that requires the careful handling of a lot of wires and data capture equipment. However, full field capture is possible and can help when dealing with anisotropic materials and structures. When choosing the method for measuring strain, it is important to consider the accuracy required. For example, some validation procedures require strain to be measured to greater than 50 µε accuracy, which is not generally feasible with DIC.

References

ASTM Standard D5379-12, 2012. *Standard test method for shear properties of composite materials by the V-Notched beam method*. American Society for Testing and Materials.

Boughanem, S., Jesson, D.A., Mulheron, M.J., Smith, P.A., Eddie, C., Psomas, S. and Rimes, M., 2015. Tensile characterisation of thick sections of Engineered Cement Composite (ECC) materials. *Journal of Materials Science*, 50(2), pp.882–897.

Grédiac, M., Sur, F. and Blaysat, B., 2016. The grid method for in-plane displacement and strain measurement: a review and analysis. *Strain*, 52(3), pp. 205–243.

Huston, C., 1879. The effect of continued and progressively increasing strain upon iron. *Journal of the Franklin Institute*, 107(1), pp.41–44.

Iosipescu, N., 1967. New accurate procedure for single shear testing of metals. *Journal of Materials*, 2(3), pp.537–566.

Jesson, D.A., Le Page, B.H., Mulheron, M.J., Smith, P.A., Wallen, A., Cocks, R., Farrow, J. and Whiter, J.T., 2010. Thermally induced strains and stresses in cast iron water distribution pipes: An experimental investigation. *Journal of Water Supply: Research and Technology-Aqua*, 59(4), pp.221–229.

Jones, E.M.C. and Iadicola, M.A. (eds.), 2018, *A good practices guide for digital image correlation*. International Digital Image Correlation Society.

Matthews, S.J., 2020. *Speckle pattern control in the application of digital image correlation for detecting damage in helmets* (Doctoral dissertation, University of Surrey).

Morice, P.B. and Base, G.D., 1953. The design and use of a demountable mechanical strain gauge for concrete structures. *Magazine of Concrete Research*, 5(13), pp.37–42.

Pannier, Y., Avril, S., Rotinat, R. and Pierron, F., 2006. Identification of elasto-plastic constitutive parameters from statically undetermined tests using the virtual fields method. *Experimental Mechanics*, 46, pp.735–755.

Thomson, W., 1883. Electrical units of measurement. *Popular Lectures and Addresses*, 1(73), pp.73–136.

10

Strain Rate Dependence: Why It Matters How Fast You Test

There is more to life than increasing its speed.

– Mahatma Ghandi

10.1 Introduction

Imagine an egg. Assuming that it is fertilised, leave it under the hen that laid it (a temperature of 35–40.5°C) for approximately 21 days and it will hatch, and a newborn chick will emerge. Now take a box of similar, albeit unfertilised eggs. Place one in a pan of water boiling at 100°C for 3 minutes and you will have a delicious snack with the white almost fully cooked and the centre runny. Leave it a little longer and it will be hard-boiled. Take another egg. Place it in vinegar for one week (replacing the vinegar after the first 24 hours) and you will remove the eggshell and make the egg become somewhat rubbery. Leave one on the kitchen counter at room temperature for a couple of months, and you will have something smelly and inedible.

This, in short, presents the fundamental problem with accelerated ageing to predict long-term performance. This is not to say that accelerated ageing tests are impossible, or that they should not be performed, but rather that they need to be considered with care: chemistry and physics conspire to change the rules when we step outside of ambient conditions.

But this book is about breaking things – why should we care about ageing samples? On the one hand, one of the reasons that mechanical characterisation is undertaken is in order to ensure that components have behaved themselves, are behaving themselves, and will behave themselves. Such aged specimens are often tested to see what the residual capacity of the specimen is in order to put limits on the lifetime of the component as the materials degrade. On the other, the problems associated with accelerated ageing are also to be found when considering strain rate, and that is firmly in the court of mechanical characterisation.

DOI: 10.1201/9780367822347-10

10.2 What is Strain Rate?

Recalling Chapter 5 and the introduction of key mechanical properties, we know that strain is the measurement of change in length divided by the original length, this change in length occurring when we apply a force to the sample. Strain rate is the way in which we measure how quickly this strain occurs. Since strain is dimensionless, being a ratio of two measurements of length, the units of strain rate are simply s^{-1}, i.e., strain per second.

In a sense, this is a secondary measurement, in that it is usually easier to measure the rate of displacement, and then convert to a strain rate. For example, if we were to assume a tensile specimen of 100 mm gauge length, i.e., the distance between the portions of the specimen gripped at the top and bottom, then a displacement rate of 6 mm per min, or, would translate to a strain rate of 0.1 $mm.s^{-1}$/100 mm = 10^{-3} s^{-1}.

10.3 Why Does Strain Rate Matter?

The rate at which strain is applied matters because the mechanisms of deformation have a reaction time: the movement of alloying or contaminating atoms in the matrix, or of dislocations and other lattice defects does not happen instantaneously. If you have ever looked at a high-frame rate video of a water balloon bursting, you will see that the water stays in place whilst the balloon almost unwraps around it. The split in the balloon opens rapidly, but not instantly. The deformation of materials is similar. If they are deformed at very low rates there is plenty of time for all mechanisms to be employed, whereas if we attempt to deform the whole very quickly there is no opportunity for the material to adjust, and it is more like two people trying to move through a door at the same time. It is important to know what strain rate we are testing at because this will affect material properties, and how the material or structure will behave in a real-world situation. For example, understanding the ultimate strength of a material and how it fails under quasi-static conditions is important, but possibly irrelevant in a car crash or a similar high-strain rate situation. Understanding how a structural member might fail and the consequences of it failing can be modelled, if we have the appropriate data.

10.4 How do We Define Strain Rate?

This is a much harder question than it might at first appear. Developing the point above we can think in terms of the material's response and so we might start with a very low strain rate, which we might associate with creep and the very slow diffusion of atoms and defects through the lattice matrix. Slightly faster, but still quite low, is quasi-static testing. By definition, quasi-static testing occurs at a rate which is slow enough that at any moment in the test, the loading condition can be considered to be static. But if we parse the time frame finely enough, could that not be the case for any test? If we look at the problem in more detail then the fundamental consideration is that the effect on the specimen at that point in time should be free of inertia. In terms of the microstructures that we have been considering to this point, there should be the opportunity for dislocations and other microstructural defects to move,[1] and similarly, there should be the opportunity for 'non-native' atoms to diffuse freely through the lattice of the parent element.[2] We are in a realm very similar to that of creep, but fundamentally there is no effect from inertia, whilst the material deforms. We might expect the material to warm slightly during testing, and this will need to be accounted for.

Faster still and the material will tend to become stiffer, and there will be less opportunity for deformation to occur, the material becomes more brittle.

At the fastest strain rates, materials will see shock waves occurring that can lead to unique modes of failure. It is at this extreme that odd effects can occur, such as projectiles passing through quite brittle materials without causing catastrophic failure.

We can therefore define a hierarchy of strain rate where quasistatic < dynamic < ballistic.

We can also, perhaps, state that very low strain rates are those that are below 10^{-3} s^{-1} and ballistic are those of the order of 10^5 s^{-1} and above. Some texts will speak very confidently of specific strain rates being representative of particular behaviours, but this confidence frequently comes from specificity: certain values will be true for some metals or polymers, but will not be true for ceramics, for example.

---

[1] Noting that there need to be slip planes free for them to move, c.f. work hardening which prevents dislocations from moving.

[2] Recalling that whether alloyed or simply present as contamination, there may be atoms of other elements present in the lattice structure that are of a different size to the parent, so that they disrupt the lattice, or sit in interstitial spaces. These will move under load.

10.5 And What about Environment?

Environment can also have its part to play, and can affect the ranges of rate that were outlined in the previous section. For example, increasing the temperature will tend to an increase in the rate of diffusivity of atoms dissolved in the lattice matrix, which will make it easier for them to move through the lattice, which will make it easier for some deformation processes to occur. The chemical environment might be such that it becomes easier for certain atoms to adsorb onto the surface and absorb into the bulk, which could act to prevent other atoms from moving, leading to a stiffening of the materials, and a more brittle response overall.

10.6 Summary

Materials will show a strain rate dependence of some sort or another. This will be a function of the microstructure of the materials and their ability to deform under certain conditions. It is important to factor this in when preparing to test materials, but be careful when looking at the literature to determine what sort of strain rate you should be applying to your specimen. The reference may be very clear as to what strain rate was applied in a specific situation, but it's important to remember that not all materials will behave in the same way.

11

The Use of Statistics

There are three kinds of lies: lies, damned lies, and statistics.

– Unknown

If you torture the data long enough, it will confess anything.
– Ronald Coase (How should economists choose?, 1982)

11.1 Introduction

The first quote at the beginning of this chapter has some mystery about it, in that it is not certain who first coined the phrase. Mark Twain attributed it to Benjamin Disraeli; some think it was a favoured aphorism of Arthur Wellesley, better known as the Duke of Wellington. Other famous politicians and learned people have also been considered as the original source.[1] Whenever it was first stated, today it is misused: it is frequently used to justify fake news, in the modern idiom, or to denigrate sound science. Statistics, like any tool, can be misused either through a lack of understanding, or to deliberately obfuscate a point, but when used properly can uncover meaning within a mess of data.

The term statistics is much changed from its original meaning. As with so much to do with mathematics, the earliest example of what we would take to be statistics is to be found in Arabic writings, in this instance of the 9th century.

---

[1] In fact, the phrase may be a recasting of a comment made in a legal context about witnesses. It is usually attributed to the judge Sir William Bramwell (later Baron Bramwell of Hever). "There are three kinds of witness. Simple liars, damned liars, and experts." He's also supposed to have finished this statement by saying "And then there's brother Fred", speaking of his brother Sir Frederick Bramwell, an eminent consulting engineer. The phrase was quoted in the 26th of November 1885 edition of *Nature*. The article explains "He did not mean that the expert uttered things which he knew to be untrue, but that by the emphasis which he laid on certain statements, and by what has been defined as a highly cultivated faculty of evasion, the effect was actually worse than if he had" (Nature, 1885).

Perhaps also unsurprisingly, this early text is on cryptoanalysis. Statistics in its modern form begins in the 17th century, however, and was related to the analysis of census information, hence stat(e)-istics. The collected information was used to inform policy – but was open to interpretation (see again the endnote on experts…). This, in a microcosm, defines much of science: we can discover all sorts of things by experimentation, but what it actually *means* may well be open to interpretation. Things might be true in a particular context, but are they TRUE? This perhaps drifts a little too far into philosophy for the current purposes, but it is an important consideration when presenting results.

In research, there is a necessity to deliver results that are as accurate as possible, but there is always a limitation. For example, recalling the suggestion in Chapter 9 that we measure the original length of a sample using a rule, marked in millimetres, the best accuracy that can be achieved is likely to be to the nearest mm. Any error will be compounded when multiple measurements are required.

Consider the measurement of the area of a square that is measured to be 10 mm on a side. The measurement of a side is 10 ± 0.5 mm. The area could therefore be as small as 9.5×9.5 mm (90.25 mm^2) or as large as 10.5×10.5 mm (110.25 mm^2) – a $\pm 5\%$ error has become a $\pm 10\%$ one, but we can reduce the error, by measuring more accurately *when it is appropriate to do so*.

Fundamentally, there is only so much testing that can be undertaken. There may be time constraints, or a limited source of material available, or budgetary constraints, or the samples may be difficult to produce. For all sorts of reasons, there are many people who will be pushing to undertake a single test, which as much as possible should be avoided: a single test will give you one data point which may be completely anomalous. Some will recall early-stage science experiments when it was drummed into them that you need to have at least three datapoints to be able to produce a straight line. In practice, three data points are insufficient, because (a) in reality your line is much more likely to be a curve, or to plateau at some point, and (b) it does not account for any uncertainty in the measurement of the data, not to mention the need for repeatability, and as we have seen elsewhere in this book, there are all sorts of situations where we might expect to need many more tests.

This chapter explores some approaches that can be used to extend our understanding of properties and performance by statistical analysis of the data. However: this is not a statistics textbook, and the purpose here is to present some of the key tools that are available and the context in which they may be used. It may even be possible to apply them appropriately from the information provided here. There are several statistical tools that can be applied, and some will be more or less relevant to a particular situation. Those presented here are some of the more popular ones, but it is always worth asking "Should I apply this here? Is there a better tool that I can use to analyse these data?". The answer to these questions can only come from a review of the data, knowledge of a range of tools, and a continuing study of the literature.

A word of warning is given to us by Ronald Coase, the Nobel Prize-winning economist: one must be careful not to manipulate the data to prove your point. The methods to be applied must be considered carefully and must be justifiable. The greater the number of operations to be applied, the more likely it is that you are drifting away from the TRUTH. And finally:

CORRELATION DOES NOT IMPLY CAUSATION.

For example, the number of sociology doctorates awarded in the US between 1997 and 2009 and the number of non-commercial space launches worldwide over the same time period seem to track each other extraordinarily closely, despite one having nothing to do with the other (Vigen, 2015). When assessing data, it is important to understand the relationship between two variables, and whether there is one there at all.

11.2 The Normal Distribution

Take a coin and flip it. Barring accidents, there should only be two outcomes: obverse or reverse – heads or tails. Those two chances should be equal, 50:50. Now flip that coin 10 times. There is a heads or tails outcome each time. If the order of the outcome is important then there are 1024 (2^{10}) possible outcomes. At the extremes, there is a 1 in 1024 chance of 10 heads in a row, and a 1 in 1024 chance of 10 tails in a row. If the exact order is not of importance, then we can aggregate certain outcomes: there are, for example ten ways in which we can get nine heads and one tail, and vice versa. These aggregated possibilities can form a normal distribution: by treating each coin toss as a binary output and summing the totals, each series will produce a number between 0 and 10. The total number of possibilities is presented in Figure 11.1a.

If we fit a profile to this distribution, we can see that it has a bell-shape, that it is centred on the mean, that it is symmetric: it is a normal distribution. Figure 11.1b demonstrates this profile and indicates the standard deviation.

The standard deviation, σ^2, for any distribution is the square-root of the sum of the value of each member of the population, x_i, minus the mean value

---

[2] Not, of course, to be confused with the sigma used for strength.

of the population, squared, and this sum total divided by the size of the population, N, i.e.:

$$\sigma = \sqrt{\sum (x_i - \mu)^2 / N} \qquad (11.1)$$

It is a characteristic of a normal distribution that, no matter how broad or narrow the distribution actually is, 68% of the values in the population will

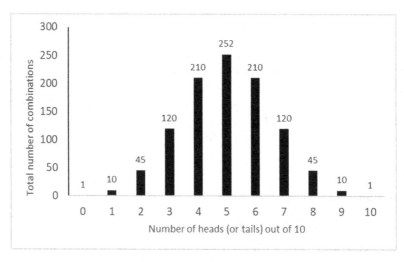

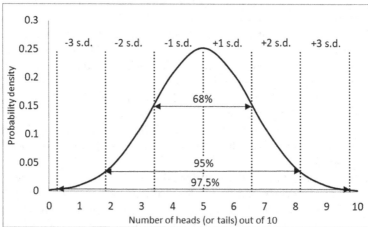

FIGURE 11.1

Possible outcomes from flipping a coin ten times in a row, presented as a normal distribution. (a) Total number of combinations for a given number of heads or tail; (b) the distribution, indicating standard deviations from the mean; (c) the distribution presented as a cumulative probability; and (d) the distribution presented as a probability density.

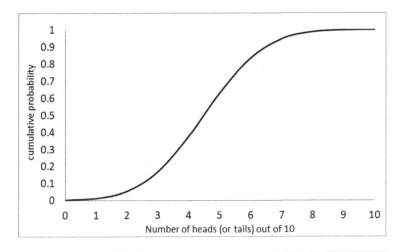

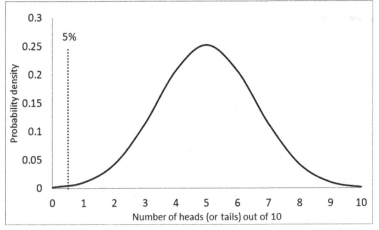

FIGURE 11.1 (Continued)

lie within ± 1 s.d., 95% within ± 2 s.d., and 97.5% within ± 3 s.d.[3] Hence, there are two things that we can take from this. Firstly, that the more that we move from the norm, the more of the population we capture. We can turn this around and say that the outliers will always be difficult to assess, because

---

[3] As an aside, this is where the six-sigma management methodology is derived from. This methodology has two guiding principles. Firstly, that any operation that gives rise to outcomes that are more than three standard deviations away from the mean need investigating. Secondly, that a company's output should come down to less than four failures in one million operations, i.e. six sigma. This could be a product, be it a loaf of bread or a car, or it could be money transaction on a website, or a phone call with someone at a support desk.

they don't turn up very often. The larger the population, the more likely it is that we will see an extreme, but we need to be careful not to give undue weight to this extreme, given the increasing number of satisfactory outcomes that were required to find this extreme outlier.

Secondly, we do need to be a little careful when applying a continuous distribution to discrete data: in the example that we are looking at the distribution implies that we could look for examples of 3.5 or 4.75 out of 10 for a series of heads or tails. This is of course impossible.

The normal distribution is important but implies that the population of data to be considered will be well behaved and give rise to something that is perfectly symmetric. This may or may not be the case in reality, and it is possible that a population of data will not follow the normal distribution. However, there are other tools.

11.3 Weibull Statistics

"A statistical distribution function of wide applicability" is a seminal paper published in 1951 by Wallodi Weibull, and it is this paper that is the cornerstone of what is today called Weibull statistics. The function truly does have a wide applicability and the examples given in the original paper include those of:[4] yield strength of a bofors steel; size distribution of fly ash; fibre length of Indian cotton; length of Cyrtoideae; fatigue life of St-37 steel; statures of men born in the British Isles; and breadth of the bean *Phaseolus vulgaris*.[5]

Whilst this first paper on the subject includes two examples relevant to materials science (and it should be noted that one is for a distribution of strengths and the other for fatigue life), it took more than 25 years for the technique to find its place, as an assessment tool for brittle solids (see, e.g., Jayatilaka and Trustrum, 1977). Here, it is demonstrated that the strength of some materials is controlled by the presence of defects. Linking this with Griffith's approach to fracture toughness, the stress required to cause a brittle material, such as a ceramic, to fail can be linked to the size of the flaw, and the probability that such defects exist in a population can be determined.

A significant advantage is that the technique can be scaled. If you can remove the defects, then you can push towards the theoretically achievable maximum strength. However, the larger the piece made, the harder it is to remove the critical defects. But if you are testing some new formulation, you

---

[4] This list as presented in the original paper.
[5] These last two presented only in an appendix in the original paper.

don't necessarily want to manufacture large-scale components, especially if your formulation is relatively rare or difficult to synthesise at the benchtop scale. In that instance, the ability to test a smaller sample and scale up the results to the desired size becomes important.

Whilst one could spend a lifetime studying Weibull statistics alone, in the context of materials, the fundamental distribution is usually described by the equation:

$$P_s(V_0) = e^{\left(-\left(\frac{\sigma - \sigma_u}{\sigma_o}\right)^m\right)}$$

(11.2)

where $P_s(V_0)$ is the probability of a sample of size V_0 surviving a stress σ. Σ_0 is referred to as the characteristic strength, a constant, and is the strength at which $P_s(V_0) = 1/e$, which is to say a strength at which 37% of the samples will survive. Aside from this, the characteristic strength has no physical significance. M is also a constant, usually called the Weibull modulus; m should always be presented as an integer value. The modulus is a measure of the variability in a sample population: a high modulus represents little variability, and a modulus of 100 or greater is usually taken to mean that within the accuracy to which the measurements are made, the strength of the material is unique. It is also worth noting that there is no physical link between m and σ_0: a material may have a high characteristic strength but display significant variability. Conversely, a material may be extremely consistent in its performance, but have a very low characteristic strength. Finally, we have the third parameter, σ_u, which is the threshold stress. This is the stress below which fracture is considered to be impossible. In some circumstances, for example in the case of ceramic materials, large flaws are entirely possible, and so the threshold stress is taken to be zero. It is also possible to dispense with this parameter when comparing two families of similar materials. If the distribution is then differentiated twice, so that the probability is presented as a double log, and the value of interest as a single log, then it is possible to present the information as a graph (Figure 11.2). Presentation can also be simplified and the reference to a critical volume can be assumed:

$$P_s = e^{\left(-\left(\frac{\sigma}{\sigma_o}\right)^m\right)}$$

(11.3)

$$\ln(\frac{1}{P_s}) = (\frac{\sigma}{\sigma_o})^m$$

(11.4)

$$\ln(\frac{1}{P_s}) = m \ln \sigma - m \ln \sigma_o$$

(11.5)

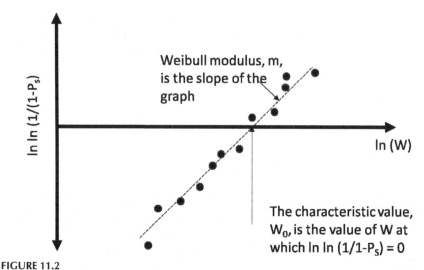

FIGURE 11.2
Illustration of a Weibull plot, demonstrating the determination of the Weibull modulus and the characteristic value.

A line of best fit through the data should yield a straight line with an R^2 correlation better than 0.9.[6] There are two things that can affect both the fit and the ultimate calculation of m and σ_0. The first is that a data set may have a slight tail to it, in terms of a small cluster of similar valued specimens. There can also be a corresponding 'neck', which is to say a small cluster of similarly high values. Under these circumstances, when presenting information, it can be acceptable to de-select some of the values in the cluster, when applying the fit, so long as this action is stated clearly. This should, however, be treated as something of a last resort and there must be a sufficiently compelling reason to do so. A second reason that the data might not yield a good fit, is that two, or potentially more, mechanisms may be in competition. For example, Smith, Mulheron, and co-workers have applied Weibull statistics to the performance of cast iron water main (Belmonte et al., 2007, 2008; Jesson et al., 2013). With respect to smaller-diameter pipe (typically of the order of 100 mm in diameter, but with examples up to 200 mm in diameter), it is possible to observe the effect of the competition between *ab initio* defects[7] and the impact of corrosion. This example also yields a useful philosophical point – just because a material has a high value for *m*, it does not mean that a material

---

[6] This is rather arbitrary, and based on personal experience, however this suggests that a correlation of 0.999 or even 0.9999 is possible with a well-behaved set of samples. Anything below 0.9 should be treated with suspicion, and everything from the samples to the experimental equipment double checked.
[7] See Chapter 4.

is necessarily 'better', simply that it is more predictable. In this instance, the presence of casting defects, and the inherent nature of cast iron arising from its microstructure mean that as-produced castings can have a considerable spread of strengths. Hence, the characteristic strength can be relatively high, but the Weibull modulus low. By contrast, the corroded metal reduces the strength of the casting, but in doing so supersedes the effect of the pre-existing defects. As a consequence, the modulus increases, but the characteristic strength is shifted down. This work characterised both pipes and small beam specimens extracted from pipes and was able to show that an assessment of a pipe taken from a particular location was able to predict the condition of a run of pipe 50 m either side of that location. This methodology was also applied to pipes that failed in service, and it was found that the characteristics of pipes from three different locations, which corresponded to three different times of installation, were very different. The older pipes, which tended to be horizontally cast had the lowest Weibull moduli, the middle batch were predominantly vertical cast, and the youngest batch, which also had the highest Weibull moduli, were spun cast. Unsurprisingly, improvements in casting processes led to pipes with fewer casting defects and more uniform microstructure (with finer graphite flakes), which consequently led to improvements in strength and a reduction in the spread of strength values. Further, the investigation also considered a series of failures, over a number of years, linked to extended periods of unusually cold weather. In these instances, it was noted that the variations in population performance were greater in the oldest sections of the network, and the more recent portions were uniformly better behaved (Belmonte et al., 2008).

It is generally agreed that there is sufficient uncertainty in the system that it is not sensible to attempt to be more precise. Another feature is that where $m \geq 100$, then the characteristic value is taken to be unique. So, for example, if one were to take 100 dog-bone samples of a particular material, and when the strength results were plotted as a Weibull distribution, if $m = 100$, then the characteristic strength would be considered to be the strength of the material. In the case of ceramics, however, the Breviary Technical Ceramics[8] suggests that most modern ceramics lie in a range $10 < m < 20$; this represents a significant improvement over the last ~40 years. This variability, due to the intrinsic defects arising from manufacturing processes that are difficult to control adequately, is one of the major barriers to significant uptake of technical ceramics.

This appreciation of m is more than a rule of thumb, but it is worth noting that even apparently 'low' values, e.g., $m = 60$, represent a reasonably narrow distribution of strengths. Figure 11.2 presents an example of a Weibull plot.

---

[8] www.keramverband.de/brevier_engl/5/3/3/5_3_3_4.htm, accessed 02/04/2020.

11.4 Monte Carlo Simulations

The Monte Carlo method was originally developed by nuclear physicists working at Los Alamos in the years just after WWII; Monte Carlo was the code name chosen for the project, and it carried through when the method was eventually released to the world.

At its least sophisticated, Monte Carlo methods can be considered as something of a brute force (in computing terms) method of considering the various permutations that could arise from the combination of a number of variables. For instance, whilst we can easily predict the result of rolling a single six-sided dice, and can work out the likelihood of rolling a total of six on two dice, what if we wanted to determine the total number of possibilities and the probability of a particular outcome for a greater number of dice? Or what if the situation were more complicated and there was a mixture of different kinds of dice, with different numbers of sides?

Monte Carlo methods are applied more or less regularly in all branches of engineering and science, both physical and social. Of particular relevance to this book is the application to reliability engineering and determining the probability of failure of complex structures such as power stations. From the perspective of understanding materials properties, Garrett (2021) investigated the mechanical properties of individual wood cells and made predictions about the overall range of behaviour by using a Monte Carlo simulation informed by variables such as the length of the cell, the height and breadth of the cell, and the wall thickness. Other, recent examples of Monte Carlo simulations include:

- Determining mechanical properties of polymer nanocomposites (Spanos and Kontsos, 2008).

- Determining the properties of fibres of spun CNTs (Wei et al., 2015).

- Investigating the effects of fibre orientation on the tensile properties of ultra-high performance fibre-reinforced concrete (Zhang et al., 2022).

A frequently cited disadvantage of Monte Carlo simulations is that they need significant computing power, although it is suggested that networked computers and other parallel computing methods can be utilised to reduce what can be a significant computing cost. However, whilst simulations are, almost by definition, numerically intense, modern computers are capable of undertaking even quite sophisticated analyses, and simulations can be prepared using languages such as Python. Simple simulations can even be built and run in Excel/VBA.

11.5 How Many Samples do I Need to Test?

When undertaking a programme of testing, it's tempting to think that we can just test one sample and then we'll have the properties that we've been looking for. In the current context we'll have a load–displacement graph that can be converted to stress–strain, we can determine the Young's modulus, and so on. But different materials will behave in different ways. Some materials are inherently 'statistical'; materials such as composites, for example, can be affected by a number of production issues. If we were to take a representative sample, say a panel 300 × 300 mm, and take suitably sized samples from this for testing, there are some materials where the test curves would essentially sit on top of one another, and others where there is some degree of variability both in terms of ultimate strength and in terms of stiffness.

The number of samples required for testing is a simple enough question to ask, but one that is difficult to answer in practice. In one sense, there are no rules to say how many samples need to be tested. But, returning to our hypothetical materials above, we will need relatively few samples of the consistent material to give us confidence in the properties, whilst a more variable performance might require a significant number of tests to ensure that we have adequately mapped the variation in properties observed.

Further, different kinds of tests might require different numbers of specimens. Some tests such as Charpy and Izod pendulum impact tests are relatively quick and easy to perform but inherently less precise and hence a larger number of tests are required to characterise a material in comparison to, say, a compact tension (CT) test or a single edge notch bend (SENB) test. Both the CT and SENB tests require greater effort to produce the specimens, carry out the tests, and analyse the results, but they are more accurate and fewer specimens are required.

On the whole though, it is easy to see that the smaller the number of tests carried out, the less credibility that the result determined has. In 'The Curious Incident of the Dog in the Night-time' (Haddon, 2003), the protagonist tells a joke to the detriment of those who make assumptions about a general population on the basis of limited data.[9] The trick with any population is to determine meaningful information without testing every single example and without losing information in the process. It is also important to ensure that inaccurate conclusions are not drawn from the data by cherry-picking, or by processing the results of testing in a manner that will introduce or reinforce bias.

Consider a selection box of sweets (see e.g., Hinde, 2020). There are 12 kinds, and 85 sweets in the box. We might hope there would be seven of each

---

[9] Précised: *An economist, a logician, and a mathematician are on a train. They cross a border and they see a brown cow standing in a field from the window of the train; the cow is standing parallel to the train. The economist says, 'Look, the cows here are brown.' The logician says, 'No. There are cows here of which at least one is brown.' And the mathematician says, 'No. There is at least one cow in Scotland, of which one side appears to be brown.'*

kind and, if we're lucky, the spare one will be our favourite flavour. But if we were to open 12 boxes, would we see one box having the 'spare' for each flavour? No – because there were not an equal number to begin with. To start, the sweets are not perfectly distributed between boxes, and whilst some quality control is conducted to provide some sort of parity, not all flavours are equally represented in the mix to begin with. Some flavours are more expensive/difficult to make, and so fewer are produced: there is already an inherent bias in the distribution. Depending on how the sweets are brought together, there may also be mixing issues, in that some sweets may slip to the bottom or rise to the top of a mix and so there are more or less of a particular kind depending on where the box falls in the packing order.

If we wanted to investigate this more thoroughly, we would need to formalise a hypothesis, perhaps suggesting that there is an absolute upper limit to the number of particular sweets in the tin, or perhaps suggesting that there are only ever a certain number, or, taking the wish above, that on average there will be seven of each type of sweet. How many tins would we need to open to prove or disprove the hypothesis?

First, we need to develop our hypothesis, and then determine the null hypothesis (H_0) that we wish to test.

If we assume that all the boxes have, on average, seven of a particular kind of sweet, this would mean that some have more, and some less, but we are assuming that the mean = the mode = the median, i.e., that most boxes contain seven, that the distribution is a 'normal' one, so that seven represents the centre of the distribution, and hence the largest number of tins will have seven. We can only have integer values, and we're expecting a reasonably tight distribution, so we'll say that the standard deviation is 1. On that basis, if we opened 1000 tins of sweets, we'd expect to get 400 tins with exactly seven sweets, 240 tins with six sweets, and another 240 with eight. There will be 55 each with five and nine, and another five each with four and ten. In theory, that is; in practice, we might need to open more to get these proportions to come out to this level of accuracy and there is a certain amount of rounding going on here. Properly there is a non-zero probability of getting a box containing any other number between zero and fourteen, but we can perhaps assume that the tails, representing chances of 1 in 10,000 or greater, are eliminated by the quality control process.

But what if we made all the same assumptions above, but we believe that the mean is actually four? The same spread, so our lowest possibility is one, and the highest is now seven. Given the numbers of tins that are sold every year, if we opened any one tin, we wouldn't be able to tell what kind of distribution it came from. So how many tins would we need to open to make a determination? This is where 'power' comes in. Variants turn up in many places, so you may have come across this elsewhere already, in another guise.

Let us break this down into the steps to be taken.

Firstly, there are two opposing theories: the probability that there are seven of a particular kind of sweet in the tin is either 40% or 0.5%. The difference here is approximately two orders of magnitude, so even if we were a few percentage points out, we could prove our point by saying that if $40 \pm 5\%$ of the tins opened have seven sweets, we can accept the null hypothesis that the mean number of a given sweet is seven.

The standard error of the mean is $\sqrt{(p(1 - p)/n)}$, i.e., in this instance $\sqrt{((0.6*0.4)/n)}$ or $0.49/\sqrt{n}$. So, we would need $0.49/\sqrt{n} \leq 0.05$, or $n \geq 96$. We'd probably round that up to 100, which is nice and neat, and just happens to be the n required for $p = 0.5$, which is a common default when p is unknown and not possible to determine. So, the good news is, we've reduced our 1000 tins to 1/10th of that.

Unfortunately, it is not quite that simple. Even if we look at 100 tins and generate a normal(ish) distribution, there is a possibility that our sampling will push us to one extreme or another – assuming our seven sweets example, perhaps all the randomly selected tins lie above or below the mean that we're trying to test. We therefore need to undertake a 'power' calculation to determine how many tins we need to test to ensure that we are testing a representative sample of the population. The sample size that we calculated above represents a power of 50%, which is to say that 50% of the possible normal distributions that could be generated from this will sit across the mean that we are testing for. The corollary of course is that 50% don't. We'd like to have a better than average chance of testing this, and this is where a power calculation comes in. How many tins would we need to test to ensure that we have a power of 80%?

To account for 95% of the distribution, we need to encompass $+/- 1.96$ s.d. of the mean. To give a power of 80% we need to determine the 80th percentile of the standard distribution, which is given by $1.96 + 0.84 = 2.8$.

From this, we can determine that we need to test 196 tins in order to have an 80% level confidence in the hypothesis that there is a specific (average) number of a particular type of sweet in a mixed tin.

In summary, looking at a tin of sweets gives us a distribution of the different kinds of sweets in the tin. We're surprised that one kind of sweet is significantly under-represented. Have we got an outlier, or is this the norm? Testing 1000 tins should give us the exact proportions of a standard distribution, but that's a lot of tins to open. Examining 96 tins would allow us to test within the standard error, but would only have a power of 50%. Typically then you would test 196 tins to give a power of 80%.

This is perhaps an extreme example. When determining the specimens to test for validation as per the Rouchon Pyramid (Figure 2.1), then we might test several thousand specimens. In a more limited piece of research, to give an indication of the properties of a material, and perhaps to show that a new process, treatment, or addition is having a demonstrable effect, we would probably not test as many specimens.

An industrial example of a similar approach to the power law is the assessment of the characteristic strength of hardened concrete, as it compares to the mean strength. This characteristic strength is similar as a concept to that defined in Weibull statistics, i.e., a value that is defined by the relationship of the family of data points, but in this case the characteristic strength is defined as the value such that 95% of the values assessed lie above this value, with the 5% below being considered defectives (British Standards Institute, 2023; CEN CR 13901, 2000). This is a design parameter. The term 'margin' is used to define the difference between this characteristic strength and the mean. The characteristic strength in this case sits at 1.64 standard deviations below the mean.

But there is also the question of how much material is available to test. As intimated in Chapter 2, we cannot test more than one of a large and/or unique structure, and indeed we may not be able to test that structure in a meaningful way. Instead, the approach outlined in Figure 2.1 is the Rouchon pyramid. We ensure that a material is sufficiently well understood so that predictive models are based on extremely accurate data. Of course, this is all very well when you have a material in mass production, or one that is reasonably well understood already. But what about the case where the material is novel and is not available in significant quantities? Jesson (2005) is an example of such a material: four different modified silicas, produced as nanoscale powders to be incorporated into a resin system. The nanosilicas were produced in a university laboratory and were available in quantities of the order of 10s of grams. This precludes a completely robust testing regime requiring kilograms of material, not to mention overcoming the issues with dispersing the particles in the resin system in the first place.

At some point it is necessary to settle on a number of specimens to test, and on a justification for that number: 6, 10, or more? If you can only test a single specimen, what can you do to maximise the value of that test?[10]

11.6 Summary

Statistics is as valid a part of engineering sciences as any other form of mathematical analysis, and indeed is a fundamental part of any aspect that requires a probabilistic answer. In the context of failure, there are several reasons to turn to statistics. There are many materials that are not especially homogeneous, and so failure is dependent on the probability of a particular flaw being present in a particular location. In other contexts, statistics can be used to provide the boundary conditions for the design of a structure: in some circumstances, a large, expensive structure will be built to withstand a 1 in 25, a 1 in 50, or a 1 in 100 event, which is to say an event that happens once in 25, once in 50, or once in 100 years. Such an event could be an earthquake, a storm, or a confluence of tides, for example. Whatever the event, the expectation is that there will be variations in the intensity of the event, but an extreme example of the event will happen infrequently. Using statistics, it is

---

[10] Actually, that's always a good question to ask, no matter how much material you have available for testing – there is always a cost to manufacturing a sample, whether overt or implied.

possible to provide some indication of how frequently the critical level event will occur and to design accordingly. The caveat, of course, is that such critical events do not occur to an exact schedule.

This chapter has considered three key topics, which can be applied in most circumstances. The meaningfulness of a statistical investigation, or the usefulness of going to the effort of conducting a Monte Carlo simulation, for example, will be dependent on the study being undertaken.

In certain circumstances, it may be necessary to look to a particular model that has been developed to deal with those circumstances. For example, in ballistics, V50 is typically used, which is to say the strike velocity at which a given threat projectile has a 50% chance of penetrating a particular target material (NATO Standardization Agency, 2015). Other verification processes by statistics are also used, and the practitioner will need to study the available tools and apply the most appropriate one to the specific setting. For example, Gumbel statistics, sometimes called the log-Weibull distribution, is popular in civil engineering, where it is used for investigating extreme value distributions (see, e.g., Zhang et al., 2014).

It is also worth noting that it may be difficult to collect enough samples to conduct an analysis to the level required in other settings. A novel material, for example, might be available in gram quantities and testing is required to justify scale-up. Under such circumstances, all one can do is conduct the tests, provide the best assessment possible, and make it clear that there are boundary conditions, and that the results observed may be overly optimistic or pessimistic.

This chapter has given only the briefest overview of statistical methods. As noted at the beginning, it is all too possible to draw unreasonable significance from data that have been overly analysed – not necessarily deliberately manipulated through the inclusion or exclusion of particular records, but just organised and reorganised and put through various operations. Raw data are seldom helpful or meaningful and need some level of tidying – much like an animal being readied for a competition: usually it will look a little scruffy, but when being judged it will look immaculate. We want people to see the significance we see: there is a difference though between, say giving a dog's tail a good brush and surgically replacing it with one that is more 'waggy'. One way of giving it 'a good brush' would be to look for some statistical significance, which is very different to adding to, or replacing some of, the data set.

One final note to conclude this chapter on statistical analysis: there are many tools available, and the choice of tool may come down to personal experience, specific training or the direction of a mentor, or exposure to the tool through books such as this one. Whilst designed for a different audience, the 'Statistical Package for Social Sciences', SPSS, is also of use and can usually be accessed for free, although do be careful to check the provenance before downloading.

References

Belmonte, H.M.S., Mulheron, M. and Smith, P.A., 2007. Weibull analysis, extrapolations and implications for condition assessment of cast iron water mains. *Fatigue & Fracture of Engineering Materials & Structures*, 30(10), pp.964–990.

Belmonte, H.M.S., Mulheron, M., Smith, P.A., Ham, A., Wescombe, K. and Whiter, J., 2008. Weibull-based methodology for condition assessment of cast iron water mains and its application. *Fatigue & Fracture of Engineering Materials & Structures*, 31(5), pp.370–385.

British Standards Institution, 2023. *Concrete. Complementary British Standard to BS EN 206: Method of Specifying and Guidance for the Specifier*. British Standards Institution.

CEN CR 13901, 2000. *The use of the concept of concrete families for the production and conformity control of concrete*; CEN Technical Report; National Standards Authority of Ireland: Dublin, Ireland.

Garrett, S., 2021. *Aspects of the Mechanical Properties of Wood Fibres for Engineered Wood Products* (Doctoral dissertation, University of Surrey).

Goldacre, B., 2014. *I think you'll find it's a bit more complicated than that*. Fourth Estate, London.

Haddon, M., 2003. *The curious incident of the dog in the Night-time*. Vintage Contemporaries, New York.

Hinde, N, 2020. *This Man's audit of quality street chocolates has the internet shook*. Huffington Post.[11]

Huff, D., 1993. *How to lie with statistics*. WW Norton & Company.

Jayatilaka, A.D.S. and Trustrum, K., 1977. Statistical approach to brittle fracture. *Journal of Materials Science*, 12(7), pp.1426–1430.

Jesson, D.A., 2005. *The interaction of nano-composite particles with a polyester resin and the effect on mechanical properties*. University of Surrey (United Kingdom).

Jesson, D.A., Mohebbi, H., Farrow, J., Mulheron, M.J. and Smith, P.A., 2013. On the condition assessment of cast iron trunk main: The effect of microstructure and in-service graphitization on mechanical properties in flexure. *Materials Science and Engineering: A*, 576, pp.192–201

NATO Standarization Agency, 2015. AEP-2920 Procedures for the evaluation and classification of personal armour, bullet and fragmentation threats. Allied Engineering Publication. Brussels: NATO.

Nature, 1885. The whole duty of a chemist. *Nature*, 33, pp.73–77.

Spanos, P.D. and Kontsos, A., 2008. A multiscale Monte Carlo finite element method for determining mechanical properties of polymer nanocomposites. *Probabilistic Engineering Mechanics*, 23(4), pp.456–470.

Steele, J.M., 2005. Darrell Huff and fifty years of how to lie with statistics. *Statistical Science*, 20(3), 205–209.

Vigen, T., 2015. *Spurious correlations*. Hachette books.

Wei, X., Ford, M., Soler-Crespo, R.A. and Espinosa, H.D., 2015. A new Monte Carlo model for predicting the mechanical properties of fiber yarns. *Journal of the Mechanics and Physics of Solids*, 84, pp.325–335.

---

[11]   Short form link: https://tinyurl.com/33p5vjcf; accessed 12/02/2021.

Weibull, W., 1951. A statistical distribution function of wide applicability. *Journal of Applied Mechanics*, 18(3), pp.293–297.

Zhang, H., Huang, Y.J., Lin, M. and Yang, Z.J., 2022. Effects of fibre orientation on tensile properties of ultra high performance fibre reinforced concrete based on meso-scale Monte Carlo simulations. *Composite Structures*, 287, pp.115331.

Zhang, W., Zhou, B., Gu, X. and Dai, H., 2014. Probability distribution model for cross-sectional area of corroded reinforcing steel bars. *Journal of Materials in Civil Engineering*, 26(5), pp.822–832.

12

Models vs. Reality

"*A theory has only the alternative of being right or wrong.*
A model has a third possibility: it might be right, but irrelevant."

– Manfred Eigen

"*Remember, all models are wrong; the practical question*
is how wrong do they have to be to not be useful."

– George Box

12.1 Introduction

As has been noted in several places in this book, physical testing of any kind is expensive. One must have equipment to do the test, material to be tested, and frequently the material must be shaped or in some other way made fit for testing. Then, as noted in the previous chapter, it is not sufficient to test just one specimen, there is the need to test a reasonable number and calculate an average and the likelihood that this average is meaningful. Further, the more complex a condition to be tested, the more complex the testing set-up needs to be; see, for example, Hafiz et al.'s (2010) work on mixed-mode testing of adhesives, and Yussof et al.'s (2017) studies on glass facades.

Models can therefore be incredibly useful. They allow us to explore situations where we cannot otherwise get information, such as in very high-speed tests where the event is so transient that it cannot be recorded by conventional means. They allow us to predict what will happen under certain conditions, so that in the case of an expensive test, such as the testing of the containment of an aero-engine in the event of the failure of a fan-blade, only one iteration is required. They allow us to repeat experiments and change the variables from the comfort of our desks, iterating to a point where we can make a single test, that will hopefully vindicate the model that has been produced. The danger is when we become too comfortable at our desks and fail to set foot in the lab. It is all too easy for students, who do not usually spend a lot of time undertaking physical testing, to believe that the results of

DOI: 10.1201/9780367822347-12

the physical test are incorrect because they don't match the theory. It is said that a modeller always believes their models when no one else does, whilst an experimentalist never believes their results but everyone else does. A lot rides on a physical model, and an experimentalist will ask all sorts of (awkward) questions about the assumptions that underpin the model. Life is very much better when experimentalists and modellers work together and learn from each other. Both are invaluable and have their place: the model can have no validity without experimental data to inform it, and experimental testing to validate it, but the experimentalist can save a lot of time and expense by targeting their experiments to the gaps highlighted by meaningful modelling.

12.2 The Worldwide Failure Exercise

The majority of materials are relatively easy to model. They behave in an isotropic manner, which is to say that if you were to take a representative piece of material, it would present the same mechanical properties no matter which direction it was tested in. At worst, it may have a small degree of directionality that can be easily understood. Composite materials are at best quasi-isotropic: it is possible to produce a lay-up that has close to the same properties in all directions, but only in a single plane. Once through-thickness properties need to be considered, i.e., the direction normal to the xy plane describing the surface, things once again become difficult.

Composites are rapidly becoming the material of choice for a number of sectors: Boeing and Airbus both have planes where the majority is formed of composite materials: the Boeing 787 Dreamliner is 50% by weight, and the Airbus A350 XWB is 53% by weight. The growth of the renewables sector relating to wind turbines, both in the size of individual turbines, and the number manufactured, is driven in part by composite materials. (Of note, the largest wind turbine blades are a similar length to an A380 aircraft, which is 73 m nose to tail.)

There are a range of constitutive models available for predicting the properties of composite materials, but designing with them in mind is challenging. Most design processes continue to treat composites as 'black metal' and fail to consider the difference in character[1] between the two materials (Spendley, 2012). As a result, some composite structures are relatively overweight and could be made more elegant if the material were understood better, whilst

---

[1] Emphasis here on 'difference': each material will have its positive and negative aspects, but overall it is important to understand that the materials are not directly interchangeable. It is necessary to understand all facets of the material, including those that might not be important until later in the life-cycle.

at the same time structures made from composites fail unexpectedly and the material is blamed when it was a design error that caused the problem right back at the beginning of things.

The Worldwide Failure Exercise has attempted to assess the viability of a range of composites failure models (Hinton, Kaddour, and co-workers, 2004, 2012, 2013a,b). It has been through three rounds, with each successive round looking at more complicated loading conditions and more involved composite layups. In all three cases, physical test data were collected and used for comparison with the model data produced using the specifications provided. The disturbing conclusion is that there are no models that perfectly predict the failure of composite materials, even in relatively simple cases, although there are some that are better than others. It is therefore necessary to be conservative when working with composite materials, more so than in other sectors – however, the point remains that it is possible to do more with less, if you understand the physical properties that you are working with. In this instance, a model can give insight, potentially, but cannot give a perfect solution.

12.3 Finite Element Analysis

Finite element methods (FEMs) are the bedrock of modelling the performance of structures. Whilst computer-aided design (CAD) is essential for producing the plans of what the thing will look like, without some way to assess what is going on, you would have to undertake a physical test. This might be difficult, expensive, time-consuming – or all three – and if the result is a failure, it is literally back to the drawing board and try again. So, if we can test the structure in question without physically testing it, we can make appropriate changes to the design and refine things until the design is, hopefully, perfect. Then we can do the difficult, expensive, time-consuming test once, to prove that the thing performs correctly.

FEMs come to us from a number of sources. The family tree, as it were, is complicated and there are several starting points, with different approaches. Some approaches pre-date electronic computers, including Hrennikoff's (1941) and Courant's (1943). However, the key point is that a large number of partial differential equations are solved for discrete regions of an object of interest. Areas which are relatively unaffected, or where generalized behaviour is expected, can be left fairly large, whilst more complicated behaviour, or regions which will see significant changes over comparatively short distances, will need elements that are small.

FEMs today are extremely powerful. Simple analyses can be performed in a matter of minutes on a standard desktop computer running freeware. More complicated scenarios will of course require more sophisticated software and more runtime.

As with any computational approach, the more sophisticated the model, the more run time and/or computational power that is required. FE software is now so ubiquitous that anyone can download a free version of reasonable capability on to a standard home PC and achieve meaningful results, although more robust, nuanced, or precise modelling will require a more powerful computer running better software, albeit that this capacity comes at a price. However, it is important to check what assumptions the software is using and to ensure that these match the reality that is being modelled. For example, if there are several possible failure criteria, is the one that you have the correct one, or just the default setting. Similarly, if the wrong relationships are used to describe the behaviour of interest, then the responses will be at best inaccurate and at worst wildly misleading. Most metals are fairly simple to model as they usually behave in an isotropic manner, whilst composites will tend to be computationally expensive to model as they are anisotropic and the mesh of finite elements needs to consider the behaviour of the fibre in particular orientations as well as the behaviour of the matrix, and will then need to consider the interface between different layers.

12.4 Physical Models

Not all models are computer generated. Scale models are often produced to demonstrate proof of concept or simply to provide a visual representation to help convey the plan in an accessible way. However, there are limits to how useful this can be, especially if the hope is to get meaningful data that can be used to inform virtual models, or further design iterations. Hence, if one were to consider producing a scale model of a material there would need to be some consideration of the microstructure of the material and how this might be controlled to produce the same responses as might be expected in a full-size specimen. For example, if one were to undertake a pull-out test of rebar from concrete at scale, one would not only need to reduce the size of the rebar, but also the aggregate and particulate size so that the microstructural length scales of the concrete microconstituents were in proportion to the rebar. Similarly, one might be able to control the processing parameters of cast iron to produce proportionately sized flakes. There is a limit to how far such behaviour can be controlled, and the meaningfulness of results, but some thought on a specific situation is always worthwhile: even if it is not possible to produce a scale model of the material, the understanding of the material will be useful in other situations.

Understanding the Failure of Materials and Structures

12.5 Summary

Whilst the focus of this book is very much on the issues which face the experimentalist and the steps that need to be taken to ensure that a mechanical test is carried out in an appropriate way, and that the maximum value is gleaned from each physical test, the ethos is that of ensuring good relations between models and reality. Physical tests are expensive, and computational power is becoming increasingly cheap. With a good understanding of the material's behaviour, robust models can be developed so that one can minimize the need for the more difficult tests, and can inform the experimentalist of the critical tests to be undertaken. Going the other way, fundamental mechanical testing can provide good-quality data with which to inform models, ensuring that the models more closely relate to reality.

References

Courant, R., 1943. "Variational methods for the solution of problems of equilibrium and vibrations". *Bulletin of the American Mathematical Society*, 49, pp.1–23.

Hafiz, T.A., Wahab, M.A., Crocombe, A.D. and Smith, P.A., 2010. "Mixed-mode fracture of adhesively bonded metallic joints under quasi-static loading". *Engineering Fracture Mechanics*, 77, pp.3434–3445.

Hinton, M.J., Soden, P.D. and Kaddour, A.S. eds., 2004. *Failure criteria in fibre reinforced polymer composites: The world-wide failure exercise.* Elsevier.

Hinton, M.J. and Kaddour, A.S., 2012. "The background to the second world-wide failure exercise. *Journal of Composite Materials*, 46, pp.2283–2294.

Hinton, M.J. and Kaddour, A.S., 2013a. The background to part b of the second world-wide failure exercise: Evaluation of theories for predicting failure in polymer composite laminates under three-dimensional states of stress. *Journal of Composite Materials*, 47, pp.643–652.

Hinton, M.J. and Kaddour, A.S., 2013b. "Triaxial test results for fibre-reinforced composites: The second world-wide failure exercise benchmark data". *Journal of Composite Materials*, 47, pp.653–678.

Hrennikoff, A., 1941. "Solution of problems of elasticity by the framework method", *Journal of Applied Mechanics*, 8, pp.169–175.

Kaddour, A.S., Hinton, M.J., Smith, P.A. and Li, S., 2013a. "The background to the third world-wide failure exercise". *Journal of Composite Materials*, 47, pp.2417–2426.

Kaddour, A.S., Hinton, M.J., Smith, P.A. and Li, S., 2013b. "A comparison between the predictive capability of matrix cracking, damage and failure criteria for fibre reinforced composite laminates: Part A of the third world-wide failure exercise". *Journal of Composite Materials*, 47, pp.2749–2779.

Soden, P.D., Kaddour, A.S. and Hinton, M.J., 2004. "Recommendations for designers and researchers resulting from the world-wide failure exercise", In *Failure Criteria in Fibre-Reinforced-Polymer Composites* (pp. 1223–1251). Elsevier.

Spendley, P.R., 2012. *Design allowables for composite aerospace structures*, Doctoral dissertation, University of Surrey.

Yussof, M.M., Parke, G. and Kamarudin, M.K., 2017. "Glass stiffness contribution of flat and curved cable-net supported glass Façade systems". *IPTEK Journal of Proceedings Series*, 3, 531–536.

13

Biomechanics

"We have the technology. We can rebuild him."

– Oscar Goldman ("The $6 Million Dollar Man")

13.1 Introduction

The $6 Million Dollar Man is an example of the interaction between medicine and engineering: whilst essentially science fiction in an otherwise real-life setting, it details the creation of a cyborg from the badly damaged body of test pilot Steve Austen. The story is almost a straight secret agent one however, in that there is no angst or conflict between the organic body, particularly the mind, and the cybernetic implants and components. This is in contrast to RoboCop, for example, which can be described as a modern incarnation of the tale of Frankenstein's monster. In either case, the implants, which are designed to augment the physical capabilities of a human being, are beyond what can currently be achieved. Whilst we have moved a long way from arguably the most famous prosthetics, the hooks and peg-legs beloved of pirates, there are still significant limits on what can be achieved.

But people have been using prosthetics for a lot longer than you might imagine. Vanderwerker (1976) provides 'A Brief Review of the History of Amputations and Prostheses', and states that the first recorded example is that of Vishpala, and is recorded in the Rig Veda.[1] This has all the hallmarks of a legend, but there are other accounts, slightly more contemporaneous but not from eye-witnesses, which include Herodotus' story of Hegesistratus (who apparently cut off his own foot in order to escape the Spartans, and replaced it with a wooden one, circa 5th century BCE) and Pliny's description of the

---

[1] It should be noted that there is some dispute over the interpretation, and also that whilst the story is dated to at least 1200 BCE, the written version of the story dates to at least 900 years after this. In all probability the events depicted are completely mythic, and it is difficult to know exactly how much truth underpins the story. From our perspective it doesn't really matter.

DOI: 10.1201/9780367822347-13

Roman general Marcus Sergius (who lost his right hand in battle during the second Punic Wars, 218–201 BCE, and had an iron one in its place, to which his shield was strapped). However, more tangible examples, for which there is archaeological evidence, include the discovery of an Egyptian mummy with a prosthetic toe dating to at least the 8th century BCE (Nerlich et al., 2000) and a (crude) prosthetic leg from around 300 BCE (e.g. Finch, 2011).

One can debate the definition of 'modern', but in general terms modern prosthetics research begins post-World War II with the work done to help the massive number of amputees arising from that conflict. Prior to this, a lot of prosthetics were custom made in their entirety, particularly with respect to the manner in which they were attached. With the increase in demand, prosthetics needed to be more robust and cheaper to produce. Whilst prosthetics need to be tailored to the individual, mass production was required to deliver the numbers demanded. But in many respects, such prosthetics were still relatively crude: compassionate innovators identified the need for prosthetics that were more like the real thing: a leg is not a solid piece, and we have articulated toes for a reason, for example. Such complexity requires the near simultaneous solution of problems relating to articulation of the joints, linkages (our muscles and ligaments), and control (replacing nerves with electronics, or somehow allowing our nerves to control the device). This is not to mention the issues around the materials to be used.

The replacement of lost limbs is also dependent on the ability to repair damaged skin and muscle tissue, however. This can be a significant issue when considering the development and use of prostheses, as the interaction between biological and non-biological moieties can be compromised by scarred tissue, which can be prone to significant chafing. There is therefore a whole area of research into the mechanical properties of skin tissue, and the effect of scarring and skin conditions on these properties, as this can affect whether or not a prosthetic can actually be used. Further to this point, there is therefore great interest in being able to define the properties of the materials used where the prosthetic interfaces with the body in order to prevent chafing (so that the prosthetic can be used for extended periods) and that the environment at this interface does not become hot and sweaty (and therefore prone to yeast infections).

So far, this Introduction has only considered prosthetics, but there are of course all sorts of other bio-mechanical solutions that require an understanding of mechanical properties to predict long-term performance, including plates to repair broken bones, hip replacements, stents to open blocked arteries – the list continues. And of course the consideration of biological systems has gone beyond just improving repairs to the body. In 1942, the prolific science fiction author Robert Heinlein published a short story in *Astounding Magazine* (under the pseudonym Anson MacDonald). The story, 'Waldo', was a classic example of fiction presenting a hither-to unthought of idea in such detail as to make it impossible for anyone to be able to claim a patent when the concept was originally developed in real life.[2] Waldo Farthingwaite-Jones is presented as a consulting engineer, who was born with *myasthenia gravis*, a neuromuscular disease, and lives on a self-designed space-station to avoid Earth's gravity. The waldoes that Waldo has designed are remote manipulators that are controlled by special gloves. The manipulators can be made for fine work, or can be used to magnify the strength of the user. The work can be done right in front of you[3] or by remote control at a significant remove.[4] Today we talk in terms of haptic devices, which is to say instrumented clothing or controllers that can provide information through the sensation of physical feedback. Pick something up, throw it, manipulate it in some kind of virtual environment and the consequences of that action are fed back to you. The haptic controllers offered for use with some games consoles are still relatively crude, but such devices are being used to create ever more realistic video games,[5] but equally are used to operate machinery in distant and/or extreme environments such as the interior of nuclear power stations or at the depths of the ocean.

This chapter explores some of the challenges that face the experimentalist when undertaking research into bio-mechanical problems, and how these can be overcome.

13.2 Canine Limbs

Biomechanics is not just about understanding the behaviour of human tissue, understanding the behaviour of animals can be important too. Some animals are too small, short-lived, or complicated to be able to fit with prostheses, but some, such as dogs can. If we think in a 'One Health' way, which is to say that there is medical research that can be carried out in a responsible, ethical manner which has immediate benefit for one population, but which can be transferable to another, the study of prostheses and the fitting of prosthetics is one such area. There is a significant body of knowledge of prostheses that has been developed for humans, which can be applied to canine care, and in adapting and developing this body of knowledge further, there is much that we can learn that can be used to benefit people.

---

[2] In fact, the story, which is considered to be a relatively minor one in Heinlein's canon, and one which adds elements of magic to an otherwise s.f. story, presents not only the mechanical waldoes, but also the waterbed.

[3] As seen in *Burning Chrome* by William Gibson. The *Encyclopedia of Science Fiction* notes that the term "is used without any further need for explanation".

[4] Something which is important to the potential success of the arcology posited in *Oath of Fealty* by Larry Niven and Jerry Pournell.

[5] Of great significance, and well ahead of current capability, in *Ready Player One*, by Ernest Cline.

In understanding how to manufacture appropriate prostheses/implants for canines, we need to understand how similar animals move when well, and the forces placed on the limb as the animal moves normally, whether walking or running. Some work can be done with living animals, such as through gait analysis, and having the animal walk across a force-plate. By matching video footage, perhaps with computer-aided analysis to map the movement, with the loading and unloading data measured on the force plate, it is possible to map the load applied with the motion.

However, there is only so far that such research can be of use. If one wishes to trial a new form of plate, or develop an understanding of how best to revise the shape of a plate, for example, then there is a need to test the implant in the appropriate conformation. It would be unethical though to trial a new device on a living creature, especially if we believed that further refinement was necessary. Instead, we can use cadaveric specimens, which can be tested in much the same way as normal mechanical test samples, albeit with some adaptations.

When attempting to mount canine cadaveric limbs for biomechanical testing, the biggest issue arises when trying to represent *in vivo* conditions. The biggest challenge in this regard is in ensuring the joint is stable with the musculature able to perform the duties they serve *in vivo*. If a test is to be meaningful then the normal motion of joints is essential, which is to say motion which prevents or limits rotation or lateral movement.

Even a brief review of the literature will show that (a) there is a lot of testing of this sort being reported and (b) there is yet to be a significant review paper which considers this kind of testing from the perspective of an experimentalist specialising in mechanical testing, rather than a veterinarian interested in the mechanical testing of prostheses. However, there are a number of features of this kind of testing that should be considered, and which can provide the focus of the development of a methodology for this kind of testing. Whilst some testing can be carried out on the tibia bone alone (the bone most likely to be damaged and need a plate), some testing requires the tibia, the tibiotarsal[6] joint, and the foot. In such a situation, there are key things to be considered. Working upwards these include:

1. The ability of the foot to move naturally.
   a. Typically sandpaper is used to manage the interface between the foot and the test-plate.[7]
   b. The tibiotarsal joint will need to be able to move.

---

[6] Effectively the ankle, although noting that in a human this would be directly above the heel which would come into contact with the ground, whereas canines essentially walk on their fingers/toes directly and transmit force through the leg to the joint, which will always be at an angle greater than 90 degrees to the floor.

[7] C.f. the question of barrelling observed in compression specimens.

2. The angle of the tibiotarsal joint. This is expected to move during normal motion, but is difficult to control during testing. Further, the angle will vary depending on the motion the animal is undertaking, e.g., walking compared to running or trotting.

3. Tendons. For a live animal, these will of course be connected up and will be helping to control mechanical behaviour. As with the question of the angle of the tibiotarsal joint, there is a question about whether the tendon is simply fixed in place or allowed to move in a controlled manner to replicate its normal behaviour. The latter is of course more difficult, but the former requires consideration of the overall pre-load that should be applied (see, e.g., Warzee et al., 2001)

4. Holding the distal end from the foot is usually effected by mounting the end of the limb in some kind of resin, but the interface between bone and 'potting' medium can be difficult to engineer, and there is a significant likelihood of failure of this interface, which can cause problems with the test (see, e.g., Wells et al., 1997).

13.3 Understanding Tissue–Tool Interactions

Medical operations are not new, but the subtlety with which surgeons can act is increasing all the time: keyhole surgery and other such techniques are now routine, and teleoperated tools, such as those described in the Introduction are not only theoretically possible, but have been demonstrated. With time to plan, many operations are minimally invasive, which improves the outcome for the patient by minimising the possibility of infection. However, the behaviour of soft tissues is still not entirely understood, and so the interaction with medical instruments does not always occur as expected.

For example, Oldfield and coworkers have investigated the problem of complex tool–tissue interactions, which can occur when needles are inserted, particularly where large strains are involved. Using phantoms, i.e., a proxy for human tissue, Leibinger et al. (2016) measured the mechanical processes of needle insertion. This study is particularly fascinating, as it also links to the consideration of the measurement of strain, as presented in Chapter 9. Here, a laser-based version of digital image correlation was used to measure displacements inside the sample as the needle was inserted. In a further piece of connectivity, this time with Chapter 12 (Models vs. reality) these data were used to inform an FE study (Oldfield et al., 2013).

13.4 Bioinspired Vibrational Control

Vibration of equipment can represent a significant hazard to health, particularly when sustained or if at certain frequencies. Damping vibration is effected by various means, but nature is providing all sorts of opportunities for new approaches. These have been summarised by Shi et al. (2021) in a review of recent research. These include studies of the way in which woodpeckers control impact as they hammer at trees, the structure of various animals' legs that enable them to move quickly over rough terrain without sustaining damage, and the necks of birds, which contain more bones than found in mammals.

13.5 Archaeological Applications

Biological specimens can also be found in history, and there is a need to understand the witness borne by marks on items uncovered. This can range from determining the force that might be imparted from an animal bite, through to the wear that might occur during the lifetime of an artefact. For example, Gargano et al. (2017) use mechanical tests as part of the characterisation of aged textiles, used for shelters in the past, whilst Calandra et al. (2020) have made the case for controlled mechanical testing experiments in archaeology, and have presented equipment for undertaking various tests to demonstrate how certain processes would develop characteristic wear patterns.

13.6 Summary

Whilst in one sense a little removed from the core focus of this book, i.e., understanding the failure of materials and structures, the testing of biological systems and their interfaces with engineering materials is important for three reasons which are directly relevant:

1. We need to understand how biological systems fail mechanically, and how they behave up until that point, for their own sake; so that we can make viable, realistic proxies; in order to repair them; and in order to make viable, realistic replacements for use when our own limbs and organs fail.

2. By understanding the behaviour of materials and structures found in nature, we can develop new materials and methods of manufacturing that can be used to improve sustainability, health, and safety, and provide innovative solutions to emerging problems.

3. By investigating archaeological specimens of a biological nature, we can gain insight into the behaviour of these materials, how they reached the condition they are in, and in some cases inform predictive models of how materials today will behave. With timber becoming a material of interest for large-scale building schemes, data from a range of sources, including archaeological ones can help with understanding the behaviour of such materials over time, even if the tendency for modern timber structures trends towards engineered wood.

References

Calandra, I., Gneisinger, W. and Marreiros, J., 2020. A versatile mechanized setup for controlled experiments in archeology. *STAR: Science & Technology of Archaeological Research*, 6(1), pp.30–40.

Finch, J., 2011, February. The ancient origins of prosthetic medicine. *The Lancet*, 377(9765), pp.548–9.

Gargano, M., Rosina, E., Monticelli, C., Zanelli, A. and Ludwig, N., 2017. Characterization of aged textile for archeological shelters through thermal, optical and mechanical tests. *Journal of Cultural Heritage*, 26, pp.36–43.

Leibinger, A., Forte, A.E., Tan, Z., Oldfield, M.J., Beyrau, F., Dini, D. and Rodriguez y Baena, F., 2016. Soft tissue phantoms for realistic needle insertion: A comparative study. *Annals of Biomedical Engineering*, 44, pp.2442–2452.

Nerlich, A. G., Zink, A., Sziemies, U. and Hagedorn, H. G., 2000, December 23/30. Ancient Egyptian Prosthesis of the Big Toe. *The Lancet*, 356, pp.2176–2179.

Oldfield, M., Dini, D., Giordano, G. and Rodriguez y Baena, F., 2013. Detailed finite element modelling of deep needle insertions into a soft tissue phantom using a cohesive approach. *Computer Methods in Biomechanics and Biomedical Engineering*, 16(5), pp.530–543.

Shi, X., Chen, T., Zhang, J., Su, B., Cong, Q. and Tian, W., 2021. A review of bioinspired vibration control technology. *Applied Sciences*, 11(22), pp.10584.

Vanderwerker, E.E., 1976. A brief review of the history of amputations and prostheses. *Inter Clinic Information Bulletin*, 15, pp.25.

Warzee CC, D.L., 2001. Effect of tibial plateau leveling on cranial and caudal tibial thrusts in canine cranial cruciate-deficient stifles: An in vitro experimental study. *Veterinary Surgery*, pp.278–286.

Wells, K.L., Pardo, A.D., Parrott, M.B. and Wassermann, J.F., 1997. A comparison of the mechanical properties of two external fixator designs for transarticular stabilization of the canine hock. *Veterinary and Comparative Orthopaedics and Traumatology*, 10(01), pp.54–59.

14

Non-destructive Evaluation

14.1 Introduction

When undertaking testing under laboratory conditions, or when undertaking a forensic examination which allows for the extraction of samples that can be examined under the microscope, or in some other way destroyed to yield an answer, the collection of key information becomes if not easy, then certainly much more straightforward.

When looking at a structure in the real world and attempting to estimate its current health and future performance, extracting samples is only feasible under very specific circumstances. Most of the time, it is necessary to collect the information required by observation, or by some non-destructive approach that may be limited in accuracy because of unknowns relating to the conformation, the microstructure, or inaccurately recorded changes that affect performance.

It would be lovely to have a handheld device that you could wave over anything that you have a vague interest in and get a read out of composition, condition, any defects, perhaps even an estimate of residual life.[1] Maybe by

---

[1] Residual life in this context refers to the service life remaining. This is a function of the likelihood of a significant event occurring, one which exceeds the capacity of the structure or component, and of the residual strength of the component which in turn is dependent on the degradation that has occurred. In this respect, then, it will be observed that the residual life may be predicted based on an assessment of the degradation that has occurred to date, and of the degradation that will continue to accumulate. Further thought will point out the two factors that mean that such predictions are helpful but should not be relied on in the long term. Firstly, obviously, the longer the period of time that the prediction extends out, the more likely it is that a catastrophic event may occur. This is compounded by the fact that further degradation will reduce the residual strength further, and therefore the required magnitude of a 'killer' event will decrease. Secondly, such predictions are based on a series of assumptions around normal operating conditions, and whilst some effort may be made to provide factors of safety, degradation can occur for different reasons, and these processes can change rate depending on conditions, which can change considerably over time. In the scenario presented here, an estimation of residual life or a prediction of future performance would simply be a case of a suite of algorithms crunching the numbers based on the measurements made by our little handheld gizmo.

DOI: 10.1201/9780367822347-14

the time the era of Star Trek rolls around, we'll be able to achieve a sensor package dense enough to achieve all the different things that are required. Or possibly not – the challenges facing such a tool are numerous and range from the complex to the mundane (see, e.g., Rainer et al., 2017). Different materials require different approaches to work out what is going on. Further, not all places are as accessible as others – our handy little tricorder can be fitted into relatively small spaces, and perhaps even poked through a hole at arm's length, where the rest of the body can't follow. Many current NDE techniques are less forgiving: they are large, unwieldy, and not necessarily suitable to be taken on-site. Frequently there is a considerable amount of setting up to do, not the quick wave of a hand – think an early photographer with their head under a hood and holding up a magnesium flash, rather than a quick snap with a mobile phone's built-in camera. Rainer et al. (2017) carried out a survey of NDE techniques and attempted to find one suitable for a particular material in a particular setting. The results were not encouraging: whilst there were a few techniques which might conceivably be able to find the defects of concern, using them on site was an impossibility.

Hence, there are two approaches to NDE. In an ideal world, we seek primary data – can we detect sub-critical defects, track them, and determine if they are growing in a manner that is likely to cause concern between now and the next survey? Alternatively, we can seek secondary data: indications that all is not well with the material that we are interested in. For example, whilst we might be able to use ultrasonics in the same way that the medical profession uses it to look at a baby in a womb, or to seek information on some internal ailments, some materials are too dense, or simply have a microstructure that precludes this. However, we can use ultrasonics in a secondary manner, driving a pulse (being careful not to introduce too much energy into the system) that can then be listened for directly, or as an echo, and the speed of sound in the material measured. If the speed is not what we expect, then we can deduce that there is a problem developing. However, some caution needs to be used when applying this technique, as not every instance of a class of material is the same, and sometimes quite large ranges for the speed of sound in a particular material can be observed: significant errors can creep if one assumes the property.

The size of the defect we are hoping to detect will be dependent on:

1. The microstructure of the material, its tolerance to cracks, and resistance to crack growth.

2. The nature of the material and the ability to interrogate it.[2] Can damage be observed directly, such as flaking paint, or surface-breaking defects,

---

[2] Glass-fibre composite materials, especially where the resin system has been matched for optical transparency, are a good choice for fundamental experiments investigating the theories underpinning behaviour, because the evolution of cracks can be observed without the need for dye penetrants or other techniques which are required for carbon-fibre composites, for example.

or internally where the parent material is transparent? Is there a technique that can detect defects in the material that we are interested in?

3. The nature of the loading in which the structure or component of interest is placed – are we expecting crack growth to be driven by creep or fatigue, or are we concerned only with the potential for unusual loading events? Will the environment play a role in crack growth?[3]

Fundamentally, there are defects that can go from sub-critical to critical in an instant, and the asset owner might not even have known that the defect was present.

14.2 Case Study: Stewarton Railway Bridge

On the 27th January 2009, a freight train pulling a mixed load of flammable liquids became derailed crossing a bridge in Ayrshire, Scotland (RAIB, 2010). The derailment was caused by the collapse of the Stewarton bridge. Ironically, the bridge was in the process of being replaced. The bridge should have been safe to use, even though one track had been taken out of use and the ballast removed. Unfortunately, the bridge was not in as good condition as had been assessed, based on decades of observations. Part of the issue was that the bridge had seen a change of duty at some point in the past (approximately 80 years before) and the structure of the bridge had been changed so that it could accommodate heavier loads. The changes included the stiffening of the structure by the addition of large plates which sealed internal portions of the bridge. Before being sealed, these portions were filled with ballast.

What went wrong? From the outside, the wrought iron structure appeared to be in good condition, and had been maintained properly, as directed by the periodic inspections. However, because the internal elements had been enclosed, these could not be observed. Water was able to penetrate into these cavities and because of the way the cavities had been enclosed and filled, water was trapped against the webs of I-beams approximately 1 m in depth. Over the course of 80 years of further use, these wrought iron I-beams had

---

[3] Consider a crack in a windscreen. The crack is sub-critical but could grow due to additional stresses arising from a period of cold weather, or through fatigue as the vehicle is being driven, or when additional load is transmitted when the vehicle gets a jolt from a speedbump or a pothole. Such a crack can be repaired relatively easily, and often an insurer will arrange for this to be done practically free of charge, rather than waiting for an inevitable claim when the windscreen is completely, or at least legally, unusable. However, we also know that whilst we can see the crack or chip, glass being good for this, there is a minimum size that will lead to a repairer being dispatched. If the crack is below this, then we just have to live with it, and keep an eye on it to see if it grows.

corroded, to the extent that the webs were little more than lace connecting the top and bottom flanges. When the train crossed the two halves of the bridge were effectively decoupled and there was counter-torque as the bridge twisted.

With hindsight, the problem is obvious, and could probably have been prevented by using a different design. With hindsight, the problem could have been detected earlier, if the right NDE tools had been used during inspections.

14.3 Structural Health Monitoring

Building on the example in the previous section, structural health monitoring (SHM) is used to understand the condition of assets, particularly those that are considered at risk, or where the consequences of failure would be severe. SHM is a four-stage process:

1. On observing the structure, is damage detected?[4]
2. On detecting damage, where is it located?[5]
3. How severe is the damage?
4. How much life remains to the structure, if nothing is done to protect it?

In the case of Stewarton bridge, the SHM process was limited to visual observation, and so the SHM process stalled at step 1. However, given the age of the structure, this is not so surprising, as no monitoring methods were applied when the structure was manufactured, or when it was repurposed. Such devices are now considered essential, especially for assets such as subsea cables, i.e., intrinsically expensive, difficult to place, and extremely difficult to monitor by normal periodic inspection methods.

14.4 Macro Digital Image Correlation

Having introduced the technique of digital image correlation (DIC) in Chapter 9, it is worth mentioning here the work of McCormick and Lord

---

[4] Here, observation is defined literally ("What do my eyes tell me about the state of this structure?") and figuratively ("What do the results of this inspection tool tell me about the condition of this structure?").

[5] This is potentially more problematic when detection arrives through a tool such as a measurement of the speed of sound, which indicates that the Young's modulus has changed, due to the presence of damage, but provides no further information.

(2012), who used macro DIC to make structural measurements across an entire bridge, monitoring the response of a bridge to various loading conditions generated by vehicles of known mass. In the same way as normal DIC identifies features that are then tracked, this macro DIC considers large structures in the same manner as small specimens and tracks features across a region of interest which just happens to be a whole building, in this paper a bridge. The authors cite the increasing accessibility of high-speed computing as fundamental to this SHM process, and this has only improved over the last 12 years.

14.5 Summary

No matter how advanced our non-destructive evaluation techniques become, it's unlikely we'll be able to eliminate all lower-rank/red-shirt fatalities in Star Trek. However, there is much that can be achieved, beyond simply looking, and not necessarily believing that what we are looking at is all that is to be evaluated. Non-destructive evaluation is the way in which we can collect evidence from structures that are in use, or which are somehow difficult to assess, although even here, care must be taken to ensure that the condition of structural elements that are hidden from sight are not forgotten when predicting future behaviour.

References

McCormick, N. and Lord, J., 2012, November. Digital image correlation for structural measurements. *Proceedings of the Institution of Civil Engineers-Civil Engineering* (Vol. 165, No. 4, pp. 185–190). Thomas Telford Ltd.

RAIB (Rail Accident Investigation Branch), 2010. *Derailment of a freight train near Stewarton*, Ayrshire, 27 January 2009.

Rainer, A., Capell, T.F., Clay-Michael, N., Demetriou, M., Evans, T.S., Jesson, D.A., Mulheron, M.J., Scudder, L. and Smith, P.A., 2017. What does NDE need to achieve for cast iron pipe networks?. *Infrastructure Asset Management*, 4(2), pp.68–82.

15

Questions for the Future

The best way to predict the future is to create it.

– Abraham Lincoln

15.1 Introduction

Following the heady days of the Space Race, humanity's forays into the wider solar system have been limited to remote probes and landers. The last person to walk on the Moon, Gene Cernan, did so in 1972. The fascination with returning to the Moon and indeed sending a crewed mission to Mars and other parts of local space where semi/permanent habitats can be built is increasing once more and has the potential to move beyond the realms of science fiction and into reality.

Back on Earth, new materials are being developed based on fungi – no longer limited to food or a threat to the integrity of a structure, fungi are finding frequent use as replacement for dense foams, such as those used for surf boards, but more recently have been developed for uses as diverse as films (Abhijith, 2018) and construction materials (Xing, 2018).

Aside from intrasolar exploration, humanity has been dealing with the extremes that nature has been presenting us with: whilst changes in averages tell one aspect of the story, they can cause us to overlook other parts. Too much of something at the wrong time is as bad or worse than too little. Excessive rain and flooding in the winter, too little rain in the summer, high winds at any time… The structures that we build are facing the upper limits of their capacity on a more regular basis. How can understanding the mechanisms of failure help us in preparing for the future?

This book has avoided exam-style questions at the end of each chapter, as are found in many textbooks: the purpose of this book is not to test the reader on what they have learned but rather to provide a guide to practical work that will be undertaken by the reader in the future, and to equip the analyst, critical friend, or simply the interested, with skills to interrogate the output of others' work and to assess the robustness of the work that has been

DOI: 10.1201/9780367822347-15

undertaken. The questions that should be foremost when considering a set of test results are therefore:

i. Do I believe these results? Are they credible with what I know of this material or structure? How can I verify these results?

ii. Is this the most appropriate test that could have been carried out? What might other tests tell us? Is the cost of further testing justified within the scenario?

iii. Given what I am looking at, should I be reviewing other instances of the use of this material? Has my understanding of how this material or structure behaves been revised?

The following sections present some thought exercises for the reader. There are no right or wrong answers to the questions posed, but in providing *an* answer the reader may be in a position to provide *the* answer, a definitive solution that will influence the area of the scenario considered going forward.

15.2 New Materials

The periodic table offers us over one hundred elements to play with, although some are more useful and stable than others. One might think that everything that is to be known about materials has already been discovered. In practice there are new ways of combining elements, new ways of processing them, and new ways of deploying them. Further, whilst the higher atomic number elements tend to be less stable and more radioactive, particle physicists postulate that there is an 'island of stability' where some super-heavy elements are stable in a manner which would allow them to be useful to materials scientists, and in the future, engineers. However:

1. The concept has been around since the 1960s (Myers and Swiatecki, 1966; Meldner, 1967);

2. Breakthroughs are only beginning to be seen now (e.g., Oganessian, 2017); and

3. Who knows how long it will be before there is enough available to be able to characterise physico-chemical properties?[1]

---

[1] Predictions can be made, and no doubt will be used to justify funding, or the denial of, projects leading to the generation and 'mass-production' of stable super-heavy elements, but the kinds of quantities required to begin testing of properties and confirm the value of potential elements will be of the order of thousands of atoms, assuming excellent isolation and recovery protocols. The generation of usable quantities really is beyond comprehension at this time.

New elements aside, there is still much ado in the world of materials, including the uptake of fungi, new methods of functionalisation leading to new products (e.g., ultrathin solar panels; Panagiotopoulos et al., 2023), and new ways of processing familiar materials (e.g. hierarchical nanostructured aluminium; Wu et al., 2019).

What opportunities might be available in the future, given the rise of a suitable new material?

Is there a danger of a new route to failure that is overlooked because we do not know enough about how the new material behaves?

How do we adapt our existing standards and characterisation processes to deal with new materials?

15.3 New Challenges

1n 1913, a comic strip began in the 'New York World' called 'Keeping up with the Joneses'. It embodied the competitive streak that consumes some people, the need to be seen as equal with, or better than, one's neighbours. Whilst this kind of competition is seen to be embodied in individuals and families, it is frequently seen in countries and cities too. Nowhere is this more evident than in architecture, and the desire for macro-entities to own, or cause to be created, or be home to, the largest of any particular kind of building. Such is the competition that various characteristics of a building or structure will be used to define whether it is the tallest, biggest, or most something. Hence, whilst Burj Khalifa, Dubai, is considered to be the tallest skyscraper in the world at 828 m, as of 2009, its highest occupiable floor is only 585 m above the ground. The rest of the structure consists of a spire, which is sometimes referred to as vanity height. Nor is it the tallest structure in the world. This epithet belongs to the Magnolia oil platform, which is 1,432 m in height, albeit that much of this is underwater.

Further, there is much competition around the use of timber in construction, which is having something of a renaissance. Large-scale wooden structures have existed for almost as long as humans have been building with timber. The 67 m tall Pagoda of Fogong Temple was built over 900 years ago and has only recently been eclipsed by the Sanctuary of Truth in Thailand, although a significant portion of its 105 m can be considered as vanity height. Similarly, if we consider the tallest wooden structures, then there have been numerous wooden radio transmitters between 150 and 190 m in height, although most of these date back to the first half of the 20th century.

As stated, timber is coming back to the fore as an important construction material for large scale. The new class of building is called a plyscraper, pun and portmanteau of plywood skyscraper. Whilst plywood is sometimes viewed as a low-grade material, particularly when compared with a nice piece of oak or teak, the advantage of plywood is the same advantage one gets in a composite material, when the fibres are aligned in different directions. The grain of the wood is now aligned in different directions, giving strength in those directions. Plywood's big brother is cross-laminated timber (CLT), and it is this that is driving the growth of mass timber construction.[2] In less than a decade, the title of tallest plyscraper has passed from Brock Common (student residence, University of British Columbia, Canada, 53 m, completed 2017), to Mjøstårnet (Brumunddal, Norway, 85.4 m, completed 2019) to the Ascent MKE Building (Milwaukee, USA, 86.56 m, completed 2022). Whilst these are modest compared with Burj Khalifa and other more traditional skyscrapers, the ambition is to create mass timber structures greater than 300 m tall within the next approximately 20 years. For example, London, UK, hopes to build a tower of ~305 m in height, and Tokyo, Japan, one of 350 m.

With these large-scale structures, how can we ensure that there are appropriate material properties available for modelling?

Can the Rouchon Pyramid provide an exemplar for the megastructures that are being built?

15.4 Out of this World

In the Introduction to this chapter the potential for Martian colonies was highlighted. If you have read Andy Weir's *The Martian*, or seen the film, you will know that a problem occurs with the habitation tent. Whilst such pods could certainly be used initially, in his Martian Trilogy, Kim Stanley Robinson posits more robust temporary accommodation and the building of larger settlements over time.[3] One proposition that is currently popular, building on emerging technology, is the '3D printing'[4] of buildings. Additive manufacturing has been increasingly used in the construction industry. The

---

[2] There are three other related materials, glue laminated timber, laminated strand lumber, and laminated veneer lumber, that are used in construction, but CLT is by far the most commonly used.

[3] These settlements frequently require the development of new materials that can be used in the Martian environment, especially where cities are built inside 'domes', i.e., tents. Others are buried deep underground.

[4] The terms 3D printing and additive manufacturing are often treated as interchangeable, but 3D printing is only one form of additive manufacturing, whereas additive manufacturing takes in some seventeen different techniques. There have been several attempts to provide an overarching standard for additive manufacturing, but these have ultimately failed, because of the complexity of identifying key features that can be found in all seventeen techniques – beyond the simple case of building a component by adding material rather than by removing material from a bulk volume.

first wire-arc additively manufactured bridge was constructed in 2017/18 (Gardner et al., 2020), and robotic placement of cement or concrete paste has been demonstrated. ISO/ASTM 52939:2023 details the requirements for such structures and is itself built on existing standards for the production and use of concrete.

There is every reason to believe that additive manufacturing of concrete buildings is a viable approach.[5] However, the building standards carry implicit bias to the conditions we find on Earth: how would these be implemented in a non-Earth atmosphere/gravity/temperature? For example, for a more traditionally constructed building based on concrete, in the UK the quality of the pour will usually be assessed by casting sample cubes at the same time as the pour and testing these after 28 days as per (standard). Will this be possible with these remotely built structures? There is also the question of long-term assessment of the structure. Our experience lies in buildings on Earth, and whilst there are a number of extreme environments that we can use to help inform our approaches, such as those that are extremely arid, there are several factors that for the moment we can only really model based on observations to date. For example, the temperature fluctuations, windspeeds, dust storms, and solar radiation (remembering the virtual lack of an atmosphere compared with Earth's) are difficult to test physically on Earth. When designing buildings, we will want to take such factors into account both from the perspective of the occupants (thermal insulation to ensure uniform internal temperature, protection from radiation, maintenance of an appropriate atmospheric pressure and oxygen content, etc.), as well as that of the building itself.

What will be the design lifetime of the planned buildings?

Will we look to extend the lifetime, or assume obsolescence?

How will we measure deterioration?

How will we determine the standards that will be used to inform the practice of observation and decision-making?

15.5 Extreme Environments

Whilst extreme environments are certainly a factor in planning for space exploration and bases on the Moon, Mars, and beyond, there are plenty of extreme environments on Earth. These can be rather special, localised situations where the temperature can rise to that of the sun, or even hotter, such as, briefly, around the path of a lightning bolt, or in the containment

---

[5] A further advantage here is that robotic printers can be sent in advance of a crewed mission and a substantial base camp prepared for the arrivals to walk straight into and occupy.

of a fusion-reactor plasma, or can be longer lasting, such as the inside of a nuclear power-station's reactor core, an Antarctic survey base, an offshore wind-farm, or the deserts of the world where water is at a premium, and temperatures can fluctuate significantly between day and night. These are all structures where failure carries environmental, commercial, and of course human, costs.

How best can we gather the information we need to help calibrate models and provide confidence in proposals to extend the operating lifetime of such structures?

What steps do we need to take to ensure the integrity of the data collected, and its meaningfulness when we come to interrogate them and incorporate them into models?

15.6 Sustainability

Sustainability is becoming a more significant factor in design and operation. The term sustainability is in danger of becoming diluted however, for reasons ranging from overuse to loose definition of what it means, to lack of understanding. Context plays an important part in understanding if something is sustainable or not. For example, buying a brand-new car every year is not sustainable for most people, but we cannot simply say that it is sustainable if the purchase comes round every ten years. It might be financially feasible, but there are other factors to consider, and then again, it might not be financially feasible because of a move to a larger house. What kind of car? How big? How much is it used?

In the context of materials, we can say definitively that there is no such thing as a sustainable material. However, we can talk about the sustainable use of materials. This requires us to think about the whole life cycle, from the extraction of the raw materials, their formation into a usable product, its disposal, and its potential for reuse. Materials from nuclear powerplants will typically need to be sequestered, so we could think in terms of the potential for carbon capture, although the embodied energy is effectively lost to the system.

In the context of failure, what can sustainability teach us, and what information do we need to be able to provide, to help people assess the sustainability of a project?

In the context of failure of materials and structures, interest is most likely to focus on the 'use phase': what do we need to do to better understand the lifetime of a structure or component?

Given that sustainability is the interaction of social, financial, and environmental concerns, how can these factors be balanced to bring about the best income?

In the context of sustainability, can failure ever be a good thing?

15.7 Closing Thoughts

Failure can be a difficult word to deal with. Socially, failure is generally considered to be a bad thing, something that brings shame on the person who has failed, and perhaps by extension on their family. This attitude may be beginning to change with the likes of the 'fail fast' attitude promulgated by the likes of Elon Musk, particularly his SpaceX venture. Move things along, fail, learn, try again, fail better, learn…

Failure is an important part of learning. My grandfather used to say that the person who had never failed, never made anything.

Failure may suffer from over-use: there have been critical failures, like those mentioned throughout this book, and there will be critical failures again – we see them played out in the news every day. But the only true failures are the failure to learn, and having learned, the failure to put into practice what we have learned.

Understanding how things fail is critical for understanding how to make them succeed, how to keep people safe, and how to change things for the better, how to help them grow.

In some respects, researchers have an easier time of failing than others: if it were guaranteed to work the first time, what would be the point? It would just be 'search'. But everyone can learn to fail, learn from failure, and succeed.

References

Abhijith, R., Ashok, A. and Rejeesh, C.R., 2018. Sustainable packaging applications from mycelium to substitute polystyrene: A review. *Materials Today: Proceedings*, 5(1), pp.2139–2145.

Gardner, L., Kyvelou, P., Herbert, G. and Buchanan, C., 2020. Testing and initial verification of the world's first metal 3D printed bridge. *Journal of Constructional Steel Research*, 172, pp.106233.

ISO/ASTM 52939:2023, Additive manufacturing for construction—Qualification principles—Structural and infrastructure elements, ISO Committee TC 261.

Meldner, H., 1967. 10. Predictions of new magic regions and masses for super heavy nuclei from calculations with realistic shell-model single-particle Hamiltonians':'. *Nuclear Chemistry Annual Report 1966*, pp.157–160.

Myers, W.D. and Swiatecki, W.J., 1966. Nuclear masses and deformations. *Nuclear Physics*, 81(1), pp.1–60

Oganessian, Y., 2017. Discovery of the Island of Stability for Super Heavy Elements. *Proc. IPAC'17*, pp.4848–4851.

Wu, G., Liu, C., Sun, L., Wang, Q., Sun, B., Han, B., Kai, J.J., Luan, J., Liu, C.T., Cao, K. and Lu, Y., 2019. Hierarchical nanostructured aluminum alloy with ultrahigh strength and large plasticity. *Nature Communications*, 10(1), pp.5099.

Xing, Y., Brewer, M., El-Gharabawy, H., Griffith, G. and Jones, P., 2018, February. Growing and testing mycelium bricks as building insulation materials. In *IOP Conference Series: Earth and Environmental Science* (Vol. 121, pp. 022032). IOP Publishing.

Index

Printed in the United States
by Baker & Taylor Publisher Services